Evolution
in
Turmoil

Evolution
In
Turmoil

An Updated Sequel to
The Troubled Waters of Evolution

By

Henry M. Morris, Ph.D.

Creation-Life Publishers
San Diego, California

Evolution in Turmoil

Copyright © 1982 Henry M. Morris

Published by
Creation-Life Publishers
P. O. Box 15666
San Diego, California 92115

ISBN 0-89051-089-X
Library of Congress Catalog Card No. 82-73600

Cataloging in Publication Data

Morris, Henry Madison, 1918 -
    Evolution in turmoil.

    1. Evolution.   2. Creation.   3. Bible and evolution.
I. Title.
A sequel to *The Troubled Waters of Evolution* by Henry
    M. Morris, published by Creation-Life Publishers,
    San Diego, in 1974.

                    575   82-73600

Cover by Colleen Dossey

Printed in the United States of America

Contents

About the Author **9**

Introduction **11**

Chapter 1 Ah, Sweet Mystery of Life **15**
No Way to Begin 15
The Vanishing Reducing Atmosphere 16
Complexity and Cosmic Bootstraps 20
Reluctant Creationists 26

Chapter 2 The Twilight of Neo-Darwinism **37**
Natural Selection in Limbo 37
Mutations and Miracles 44
Chaotic Genetics 50
Hemoglobin Hierarchies 52

Chapter 3 Where Have All the Fossils Gone? **61**
The Case of the Lost Connections 61
Men, Monkeys, and Missing Links 72
The Hopeless Monsters 83

Chapter 4 Evolution and Revolution **91**
How to Change Society 91
The Racism of Darwinism 94
Marxist Punctuationism 98
Naturalistic Catastrophism 105

Chapter 5 Anti-Creationist Hysteria **113**
The Return of Creationism 113
Why Do Creationists Win Debates? 116
The Battle for the Public School 123
The War Against Creation 135

Chapter 6 Christian Evolution and **141**
            Flaming Snowflakes
Evolution and the Compromising Christian 141
Christians and the Uncompromising Evolutionist . . 144
The Non-Christian Nature of Evolution In Action . . . 149
Did God Really Mean What He Said? 159

Chapter 7 The Way of Peace **167**
Peace or Turmoil 167
The Troubled Soul of Charles Darwin 169
The Creator of Peace 171

Indexes **177**

About
The
Author

Henry M. Morris is President of the Institute for Creation Research and has been actively working in the field of scientific Biblical creationism for almost 40 years. He is author or co-author of many widely used books on creationism (e.g., *What is Creation Science?*, *The Genesis Flood*, *Scientific Creationism*), on Biblical studies (e.g., *The Genesis Record*, *The Revelation Record*, *The Bible Has the Answer*), and in his own scientific field (e.g., *Applied Hydraulics in Engineering*). He has the B.S. from Rice University and the M.S. and Ph.D. from the University of Minnesota. He is a full member of Sigma Xi and Phi Beta Kappa, as well as a Fellow of the American Association for Advancement of Science and the American Society of Civil Engineers. His professional society memberships include the Geological Society of America, the American Geophysical Union, and others. He spent 28 years on the faculties of five important universities, including 18 years as chairman of major academic departments, before becoming co-founder of Christian Heritage College and the Institute for Creation Research in 1970. He is a former president of Christian Heritage College, as well as a founder and former president of the Creation Research Society. He and his wife, Mary Louise, now live in San Diego. They are parents of three sons and three daughters, all of whom are active and happy Christians, and to date have twelve grandchildren.

Introduction

When *The Troubled Waters of Evolution,*[1] to which this book can be considered a sequel, was published in 1975, the creation movement was just beginning to come to national attention. The first of the campus creation/evolution debates had been held in October 1972, and, by the time the book was published, about a dozen had been held on various campuses in the United States and Canada, plus a couple in New Zealand. The ICR (Institute for Creation Research, San Diego) textbook, *Scientific Creationism*[1] had been published in 1974, but no one anticipated it would eventually receive such wide attention from the scientific and educational establishments. To suggest that the sacrosanct doctrine of

---

1. These two books, *Scientific Creationism* (Ed., H. M. Morris, Creation-Life Publishers, 1974, 277 pp.), and *The Troubled Waters of Evolution* (H. M. Morris, Creation-Life Publishers, 1975, 217 pp.) were among the first developed by the Institute for Creation Research. They have exerted wide influence in the creationist movement.

evolution might be heading into troubled waters seemed unrealistic at best.

For the most part, the scientists and teachers in these establishments were still ignoring creationists and their arguments. When they noticed them at all, their reaction tended to be one of patronizing amusement.

Now, however, seven years later, creationists finally have their attention! At this writing (September 1982), their attitude is no longer one of indifference or amusement, but one of anger and hostility. As far as students and the general public are concerned, on the other hand, there has been a real upsurge of interest and acceptance. A Gallup poll in 1979, for example, found that over half the American people now believe in a literal Adam and Eve, specially created by God, in spite of decades of indoctrination in human evolution by the schools and news media. In 1981, an NBC-Associated Press poll discovered that at least 86% of the American people now desire to have creation taught in the public schools. In mid-1982, a Gallup poll found that almost half the population has even come to believe in *recent* creation! The evolutionary establishment finally has good reason to be concerned.

In this book, therefore, we want to survey these developments of the past seven years. The evolutionists are in real turmoil today, not only because of creationist pressures, but perhaps even more because of internal disagreements. The entire structure of neo-Darwinism, dominant for so long as the supposed mechanism of evolution, has all but collapsed, even though evolutionists protest that they are as committed as ever to the general concept of evolution. It has been an exciting and tumultuous seven-year period, and the years just ahead may be even more so. This seems to be an appropriate time to survey the increasingly tumultuous evolutionary sea, with its many scientific shallows and the many storms that are blowing over its troubled surface more tumultuously than ever before.

Chapter 1

Ah, Sweet Mystery of Life

NO WAY TO BEGIN

Evolutionism, of course, is supposed to be a total world view, explaining not only the origin of species, but the origin of life itself. If a transcendent God is allowed into the processes of the universe at all, then there is no real reason not to accept the total world view of creationism. If God was able to create *any* life, then it would be no great thing for Him to create *all* the basic kinds of life, each fully developed, exactly as the Scriptures say He did. The greatest gap of all is that between the living and the nonliving. If God could do *that*, then He could surely do the rest. In fact, that would be the only reasonable and merciful thing to do. Why waste time on an age-long meandering, tortuous spectacle of struggle and suffering in order to produce man? With God it certainly would not have been necessary.

Consequently, committed evolutionists have always tried to explain the origin of life on a purely naturalistic basis. Although Charles

Darwin at first allowed the possibility that a Creator may have generated the first spark of life, in his later life and writings he was a consistent evolutionist—which means an atheistic evolutionist—and the *leaders* of evolutionary thought ever since have all believed that life arose all by itself. Although many *followers* have been theistic evolutionists, the *leaders* have been either atheists, pantheists, or agnostics.

But their commitment has always had to be one of faith, *not fact.* The common belief in "spontaneous generation," once almost universally held by evolutionary scientists, is, of course, no longer held at all. The problem with this idea, like that of Lamarckianism, was that it was *testable.* As soon as it was carefully tested, especially by the creationist Louis Pasteur, it was shown to be totally false.

But that only meant—to devout evolutionists, at least—that life was not arising *now.* They could much more safely argue that it somehow arose in the past, when conditions were different. That way, it could not be disproved.

THE VANISHING REDUCING ATMOSPHERE

So, for almost a century, evolutionists have been trying to develop some kind of rational scenario as to how life might have originated when conditions were just right sometime in past aeons. The most widely accepted and influential proposal was that of the Russian Communist, biochemist Alexander Oparin, who in 1936 wrote his treatise on *The Origin and Evolution of Life,* postulating that life arose in a primordial soup of organic chemicals under energizing influences in some kind of reducing atmosphere. The latter was considered absolutely essential, since any primitive biologically active molecules which might have formed would have been immediately destroyed by oxidation in an oxidizing atmosphere such as we have at present.

Thus evolutionists propose that some unknown primitive life form arose by some unknown process in a sea of unknown chemistry in contact with an atmosphere of unknown composition bombarded with unknown energies from an unknown source! This seems to remove the whole system far beyond the realm of testability, and so makes it extremely attractive to evolutionists. This scenario accordingly has been incorporated in just about every biology textbook ever since, presented as one of the sure conclusions of modern science. The idea was strengthened by the famous 1953 Stanley Miller experiment (also now included in all the textbooks) which produced certain amino acids in a specialized apparatus supposed to model the hypothetical effects of the primeval reducing atmosphere. A recent review article by Dr. John Gribbin, in England, commented as follows:

> It used to be widely thought, and widely taught, that the original "primitive" atmosphere of the early Earth was a "reducing" atmosphere, that is with no oxygen but rich in hydrocarbons such as methane and ammonia, which can combine with oxygen . . . . The reasoning behind this assumption developed primarily from the belief that such an

atmosphere would be ideal, and might be
essential, for the development of the
complex but nonliving molecules that
preceded life . . . . This picture captured
the popular imagination, and the story of
life emerging in the seas or pools of a
planet swathed in an atmosphere of
methane and ammonia soon became part
of the scientific folklore that "every
schoolchild knows."[1]

Thus does one generation's "assured scientific
fact" become the next generation's "scientific
folklore"! The abandonment of the primeval
reducing atmosphere is only the first of the
many aspects of evolutionary dogma that are
currently being traumatically modified or
abandoned today.

The late-lamented reducing atmosphere has
so recently been dissipated, that most
evolutionists are not even aware of it.
Nevertheless, the space program, with its
studies of the atmospheres of Mars and Venus,
together with increased knowledge of the
components of volcanic gases from the earth's
interior, has just about forced its abandonment
by knowledgeable scientists.

Although biologists concerned with the
origin of life often quote an early
atmosphere consisting of reduced gases,
this stems as much from ignorance of
recent advances as from active opposition

1.   John Gribbin, "Carbon Dioxide, Ammonia, and Life,"
      New Scientist, Vol. 94, May 13, 1982, p. 413.

to them . . . . The time has come, it seems, to accept as the new orthodoxy the idea of early oxidized atmospheres on all three terrestrial planets, and the biological primers which still tell of life on Earth starting out from a methane/ammonia atmosphere energized by electric storms and solar ultraviolet need to be rewritten.[1]

Geologic evidence often presented in favor of an early anoxic atmosphere is both contentious and ambiguous . . . . Recent biological and interplanetary studies seem to favor an early oxidized atmosphere rich in CO_2 and possibly containing free molecular oxygen . . . . It is suggested that from the time of the earliest dated rocks at 3.7 b.y. ago, Earth had an oxygenic atmosphere.[2]

Think of all the textbooks that now have to be rewritten and all the millions of tax dollars that have been wasted on pointless experiments designed to show how life might have formed under these exotic planetary conditions which never existed! On the other hand, think also of all the newly revised textbooks which will now need to be purchased for new classes of school children and all the

1. Monitor, "Smaller Planets Began with Oxidized Atmosphere," *New Scientist*, Vol. 87, July 10, 1980, p. 112.
2. Harry Clemmey and Nick Badham, "Oxygen in the Precambrian Atmosphere: An Evaluation of the Geological Evidence," *Geology*, Vol. 10, March 1982, p. 141.

costly experiments that will be devised to try to elucidate how life could arise in an oxygen-abundant atmosphere.

> All we have to do now is rewrite all those textbooks and ensure that "every schoolchild knows" what the best theory of the evolution of the Earth's atmosphere and the origins of life is today: . . . .[1]

As yesterday's dogma is today's folklore, so probably today's orthodoxy will become tomorrow's outdated mythology. Evolutionary theory is always evolving (even though the *world* doesn't), "ever learning, and never able to come to the knowledge of the truth" (II Timothy 3:7). Bible-believing creationists can at least take some comfort in noting that the new primeval atmosphere is closer to the Biblical truth than the old primeval atmosphere.

COMPLEXITY AND COSMIC BOOTSTRAPS

There have been, of course, certain studies that seem to indicate the possibility of life arising in an oxidizing environment, even though this was considered impossible a few years ago. Without such hope, evolutionists would have been unwilling to give up their reducing atmosphere, no matter what the facts of planetary geology seemed to demand.

But neither type of atmosphere provides a solution to the real problem—that of the tremendously complex information system that

---

1.   John Gribbin, *op cit*, p. 416.

must somehow be acquired even by the simplest imaginary form of life to enable it to specify its own replication. As Douglas R. Hofstadter (himself a committed evolutionist) has admitted:

> A natural and fundamental question to ask on learning of these incredibly interlocking pieces of software and hardware is: "How did they ever get started in the first place?" It is truly a baffling thing. One has to imagine some sort of bootstrap process occurring, somewhat like that which is used in the development of new computer languages —but a bootstrap from simple molecules to entire cells is almost beyond one's power to imagine. There are various theories on the origin of life. They all run aground on this most central of all central questions: "How did the Genetic Code, along with the mechanisms for its translation (ribosomes and RNA molecules), originate?" For the moment, we will have to content ourselves with a sense of wonder and awe, rather than with an answer[1]

A similar admission has been made recently by one of the most distinguished scientists working in this field:

> We do not yet understand even the general features of the origin of the

---

1. Douglas R. Hofstadter, *Godel, Escher, Bach: An Eternal Golden Braid* (New York: Vintage Books, 1980), p. 548.

> genetic code . . . . The origin of the genetic
> code is the most baffling aspect of the
> problem of the origins of life, and a major
> conceptual or experimental breakthrough
> may be needed before we can make any
> substantial progress.[1]

As a matter of fact, nothing less than a miracle
would be required to create life in nonliving
chemicals. Creationists have long insisted that
the Second Law of Thermodynamics constitutes
an impregnable naturalistic barrier against any
such event ever occurring by chance. In
response, evolutionists have simply, and
monotonously, continued to berate creationists
for misrepresenting the Second Law. Thermodynamics applies only to isolated systems, they
say (although no thermodynamics textbook
says such a thing), and creationists don't
realize that the earth is an open system
(although creationists do realize this and have
always developed their arguments in the
context of open-system thermodynamics).
Evolutionists seem to have a strange mental
block at this point. If they *really* understood the
legitimate implications of the Second Law in
origin-of-life problems, of course, they would
have to repudiate either evolution or the
Second Law.

One of the few evolutionary biochemists who
has seriously addressed this problem is

---

1.   Leslie Orgel, "Darwinism at the Very Beginning of Life,"
     New Scientist, Vol. 94, April 15, 1982, p. 151.

Germany's Manfred Eigen, and he rather arrogantly slides over the problems encountered in his efforts. Note the following statements culled from a recent article trying to imagine this process.

> It was therefore necessary for the first organizing principle to be highly selective from the start. It had to tolerate an enormous overburden of small molecules that were biologically "wrong" but chemically possible . . . . The primitive soup did face an energy crisis, early life forms needed somehow to extract chemical energy from the molecules in the soup. For the story we have to tell here it is not important how they did so; some system of energy storage and delivery based on phosphates can be assumed.[1]

This requirement of an energy storage and delivery system is exactly what creationists have always insisted is necessary to drive a system upward toward higher complexity against its innate tendency (as expressed in the Second Law of Thermodynamics) to go downward. Note, however, that Eigen did not *describe* such a system—he merely *assumed* it! Further, he assumed it was somehow based on "phosphates," which are, indeed, important energy storage-and-delivery systems in

---

1.   Manfred Eigen, Wm. Gardiner, Peter Schuster, and Ruthild Winkler-Oswatitsch, "The Origin of Genetic Information," *Scientific American*, Vol. 244, April 1981, p. 88.

already living-and-replicating organisms. But these marvelous "batteries" are themselves produced by the metabolism of the living systems they energize. Eigen also glides over this problem:

> Non-metabolic replenishment of the phosphate energy reservoir . . . had to last until a mechanism evolved for fermenting some otherwise unneeded components of the soup[1]

Ah, here is another unknown *mechanism* which must be *assumed* in order to reverse the downhill direction otherwise specified by the Second Law. But that is not all.

> One can safely assume that primordial routes of synthesis and differentiation provided minute concentrations of short sequences of nucleotides that would be recognized as "correct" by the standards of today's biochemistry . . . . The primitive RNA strands that happened to have the right backbone and the right nucleotides had a second and crucial advantage. They alone were capable of stable self-replication.[2]

Exactly what these "primordial routes of synthesis and differentiation" may have been,

---

1. *Ibid.*
2. Eigen, *et al, op cit*, p. 91.

we, of course, do not know, but we may, nevertheless, "safely assume them"! And although many incorrect sequences of nucleotides no doubt were assembled in the unknown processes of this scenario, those which happened to be correct were able to reproduce themselves and so managed to survive, so the script goes. This imaginative scenario is offered to us in all seriousness as an explanation of how life could have evolved from nonlife in spite of the negative pressures imposed by the Second Law. Eigen and his associates have dressed these statements of faith up in impressive mathematical and biochemical verbiage and have even claimed experimental confirmation of one or two steps in the long process, so that many evolutionary biologists now merely refer to Eigen when confronted with the Second Law argument against evolution.

Nevertheless, the basic problem is still completely unsolved. Eigen, *et al*, did not even attempt to deal with the barrier of stereo-chemistry (the universal prevalence of "left-handed" amino acids in the protein structure of living systems, as against the uniform distribution of left-handed and right-handed molecules in nonliving systems), nor with the development of nucleic acids into proteins, nor with many other key problems in the origin of life. Furthermore, they could not solve the all-important "chicken/egg" problem.

Which came first, function or information? As we shall show, neither one

could precede the other; they had to
evolve together.[1]

Despite the high claims and reputation of their advocates, such speculations are pointless.

RELUCTANT CREATIONISTS

The distinguished British astronomer, Sir Fred Hoyle, has in recent years been stressing the fact that thermodynamics considerations, in the light of such complexities as those ignored in the treatments of Eigen and the American origin-of-life chemists, have made it impossible to believe any longer in the naturalistic origin of life on the earth.

> I don't know how long it is going to be before astronomers generally recognize that the combinatorial arrangement of not even one among the many thousands of biopolymers on which life depends could have been arrived at by natural processes here on the earth. Astronomers will have a little difficulty in understanding this because they will be assured by biologists that it is not so, the biologists having been assured in their turn by others that it is not so. The "others" are a group of persons who believe, quite openly, in mathematical miracles. They advocate the belief that tucked away in nature, outside of normal physics, there is a law which performs miracles (provided the miracles

---

1. *Ibid.*

are in the aid of biology). This curious
situation sits oddly on a profession that
for long has been dedicated to coming up
with logical explanations of biblical
miracles . . . . It is quite otherwise, however, with the modern miracle workers,
who are always to be found living in the
twilight fringes of thermodynamics.[1]

These "twilight fringes of thermodynamics"
to which Hoyle refers are exactly what
creationists have been talking about for years,
the very existence of which evolutionists
commonly ignore if they do not deny. Merely
having an energy field available to an open
system by no means implies that the system
will automatically increase in organized
complexity. Hoyle has described this absurd
conception of the evolutionist in picturesque
terms.

The chance that higher life forms have
emerged in this way is comparable with
the chance that a tornado sweeping
through a junk-yard might assemble a
Boeing 747 from the materials therein.[2]

No creationist has ever denied that the earth
is an open system (although evolutionist after
evolutionist keeps repeating this falsehood),
nor that it has always had an abundance of

---

1. Fred Hoyle, "The Big Bang in Astronomy," *New Scientist*, Vol. 92, November 19, 1981, p. 526.
2. "Hoyle on Evolution," *Nature*, Vol. 294, November 12, 1981, p. 105.

solar energy accessible to its processes. Solar energy sustains the earth's atmospheric circulation and its hydrologic cycle, but it does not generate life in an organic soup. Solar energy produces tornadoes, but tornadoes do not fabricate airplanes. Solar energy perpetually bathes the building materials resting on a construction site, and these constitute a patently open system, but it would never in a million years organize these components into a building. How can evolutionists remain so stubbornly blind to the fact that, thermodynamically, an energy field imposed on a complexly organized open system will cause the system to become *less* organized, not *more* organized, *unless* that open system also has intrinsic to its structure a pre-programmed information system (e.g., the genetic code) and a pre-packaged storage/conversion/delivery system for the incoming energy (e.g., photosynthesis) to direct and energize its growth in complexity?

The thermodynamic barrier to the naturalistic synthesis of life has often, and alternatively, been expressed in terms of prob-abilities. Hoyle, with his gift of picturesque language, has expressed it thus:

> At all events, anyone with even a nodding acquaintance with the Rubik cube will concede the near-impossibility of a solution being obtained by a blind person moving the cubic faces at random. Now imagine 10^{50} blind persons each with a scrambled Rubik cube, and try to conceive of the chance of them all *simultaneously* arriving at the solved form. You then have

the chance of arriving by random shuffling
at just one of the many biopolymers on
which life depends. The notion that not
only biopolymers but the operating
programme of a living cell could be
arrived at by chance in a primordial
organic soup here on the Earth is
evidently nonsense of a high order.[1]

But evolutionary biologists have been
confronted with this probabilistic argument
many times before, and they respond merely by
ignoring it. The only alternative to a
naturalistic origin of life is the creation of life,
and this they consider to be unthinkable. No
matter how improbable it may be, life must
have arisen by chance somehow, because here
we are! . . . so the "reasoning" goes.
Creationists well understand Hoyle's reaction to
his current treatment by his former admirers:

Of adherents of biological evolution,
Hoyle said that he was at a loss to understand "biologists' widespread compulsion
to deny what seems to me to be obvious."[2]

Many evolutionists have sought to escape the
probability argument by appealing to the great
expanse of geological time. Any impossible

---

1. Fred Hoyle, *The Big Bang in Astronomy,* p. 527. The
   number of possible scramblings of the Rubik cube is
   4×10^{19}—that is, forty billion billion. The 10^{50} blind
   persons in the illustration would be a hundred trillion
   trillion trillion trillion people.
2. "Hoyle on Evolution," p. 105.

thing becomes possible if there is enough time, they hope.

There is really not that much time, however, even if the geological ages have been correctly identified and dated by the accepted methods of geochronometry. The problem is that life originated almost as soon as the primeval earth became cool enough to permit life to survive. The earth is said to be 4.6 billion years old, but the oldest dated rocks are assigned a 3.8 billion-year date and the oldest life forms (fossil prokaryotes in South Africa) are dated at more than 3.4 billion years. As Stephen Jay Gould says:

> . . . we are left with very little time between the development of suitable conditions for life on the earth's surface and the origin of life . . . . Life apparently arose about as soon as the earth became cool enough to support it.[1]

Now here is a marvelous thing! Two of the world's leading scientists, both of whom have been outspoken agnostics (although Hoyle has recently been "converted," sort of) have approached the subject of the origin of life from two quite different perspectives. Hoyle has concluded that life could never have originated naturalistically on earth, while Gould has said that life originated on the earth almost as soon as the earth cooled enough for life.

---

1. Stephen Jay Gould, "An Early Start," *Natural History*, Vol. LXXXVII, February 1978, pp. 10, 24.

As a matter of fact, they are *both* right! Life could *never* have originated naturalistically, but it *did* appear on the earth almost as soon as the earth was created (just five days later, to be precise). None of these evolutionary speculations have ever, in the very least degree, refuted the straightforward Biblical record of how life began.

But Gould, instead of acknowledging creation, has instead used the sudden origin of life on earth as an argument supporting his Marxist approach to evolution, the so-called "punctuated equilibrium" mechanism (see Chapters 3 and 4).

> Gradualism . . . . was primarily a prejudice of nineteenth-century liberalism facing a world in revolution. But it continues to color our supposedly objective reading of life's history . . . . The history of life, as I read it, is a series of long stable states, punctuated at rare intervals by major events that occur with great rapidity and set up the next stable era . . . . My favorite metaphor is a world of occasional pulses, driving recalcitrant systems from one stable state to the next.[1]

So Gould takes the position that life arose explosively, just as did all later evolutionary advances in the history of life. The particular reaction—or series of reactions—which thus

---

1.   *Ibid,* p. 24.

quickly transmuted chemicals into living cells remains yet to be discovered (or imagined, or invented), of course.

Sir Fred Hoyle, on the other hand, says it is inconceivable thermodynamically that life could have arisen by chance here on the earth at all, even in five billion years. Instead of concluding that it therefore must have been created on the earth, Sir Fred says that life must therefore be a cosmic phenomenon, translated to earth somehow from outer space.

Some have dismissed his ideas rather smugly, since he is an astronomer and mathematician,[1] not a biochemist. However, the "directed panspermia" theory of Leslie Orgel and Francis Crick (co-discoverer of DNA) is not too much different from Hoyle's speculations, and these two men are among the world's top biochemists. They believe that "life-seeds" have been directed through space by advanced civilizations in other galaxies— hence the term *directed panspermia.* Hoyle speaks of a "life cloud" permeating space. Both agree that life could not have evolved on the earth by chance.

Hoyle and his colleague, Chandra Wick-

---

1. Sir Fred Hoyle is best known as the originator of the Steady-State Theory of cosmogony, but he has written extensively on many important scientific themes. For many years he was University Lecturer in Mathematics at Cambridge University. He is now an honorary research professor at Manchester University and University College, Cardiff.

ramasinghe, have gone even further. The same type of probability calculation which they applied to geologic time on earth was also applied, with only slightly modified numbers, to cosmic time in the whole universe. Thus they conclude finally that life in space must be the product of intelligent creation, and are even willing to speak of God.

> Once we see that the probability of life originating at random is so utterly minuscule as to make it absurd, it becomes sensible to think that the favourable properties of physics, on which life depends, are in every respect *deliberate.* It is almost inevitable that our own measure of intelligence must reflect higher intelligence—even to the limit of God.[1]

These two very distinguished scientists have thus recently become converts to creationism and have already suffered much rejection and ridicule by their erstwhile colleagues. Dr. Wickramasinghe has testified that he was previously a consistent atheistic Buddhist but has reluctantly been forced to a creationist position by the scientific facts. He was even willing to testify for the creationist side at the creation-law trial in Arkansas in 1981.

The most interesting aspect of their

---

1.   Fred Hoyle and Chandra Wickramasinghe, *Evolution from Space,* (London: Dent, 1981). Wickramasinghe is Professor of Astronomy and Applied Mathematics at University College, Cardiff.

conversion to creationism was that the Bible had nothing to do with it. Evolutionists have continually insisted that creationists are creationists because of their belief in the Bible, so the Hoyle-Wickramasinghe conversion has been a real embarrassment. These two scientists do not believe in either the Biblical record of creation or even in the God of the Bible. Nevertheless, they insist that solid scientific considerations compel them reluctantly to conclude that life can only be the product of some cosmic Intelligence, not of any kind of naturalistic evolutionary process, and they are even willing to call Him God.

Evolutionists thus still have no factual evidence, or even a workable model, to account for this very first stage in organic evolution. If there is any kind of evolutionary family tree, it is a tree without roots. Their cherished reducing environment is rapidly vanishing, even with all the elegant speculations of Oparin and the endlessly-cited experiments of Stanley Miller and Sydney Fox. These have all suddenly become irrelevant. The infinitely complex structures, symbiotic relationships, and informational programs found in all living cells are separated by an unbridgeable thermo-dynamic and probabilistic gulf from all nonliving systems, and more and more *careful* scientists are recognizing this fact today.

Why are most evolutionists still so stubbornly unwilling even to consider the possibility that life was created by the living God? Have they proved that God does not exist? Of course not, and no scientist of rational mind would ever claim that he had. Does the study of *science* exclude God by definition, as many today would allege? Not if science really involves the search for truth, as they also say. If the existence and power of God are even possibly true, then their implications are legitimate components of scientific discussion—as all the founding fathers of science would certainly have agreed.

Isaac Asimov, probably the most prolific science writer of all time, is at least honest on this point:

> Emotionally I am an atheist. I don't have the evidence to prove that God doesn't exist, but I so strongly suspect he doesn't that I don't want to waste my time.[1]

It is this kind of irrationality that governs most advocates of origin-of-life theories. But should their prejudices prevent others from considering God? The Psalmist long ago gave us his evaluation in the classic words, "The fool hath said in his heart, There is no God" (Psalm 14:1; 53:1).

---

1. Isaac Asimov, interview in *Free Inquiry*, as quoted in Context, June 15, 1982.

Chapter 2

The Twilight of Neo-Darwinism

NATURAL SELECTION IN LIMBO

For many years, creationists have been pointing out the logical fallacy involved in attributing evolution to natural selection, stressing the inherently tautologous nature of the whole concept. That is, natural selection was supposed to insure "the survival of the fittest," but the only pragmatic way to define "the fittest" is "those who survive." Thus the long neck of the giraffe and the short neck of the hippopotamus are both explicable by natural selection, as are both the dull coloration of the peppered moth and the brilliant colors of the bird of paradise. Natural selection "explains" everything, and therefore really *explains* nothing!

Creationists had posed a similar objection to the evolutionist's concept of "adaptation." The fact that a particular organism is adapted to its environment tells us nothing whatever about how it *became* adapted. Any organisms not so adapted would not have survived, but this constitutes no proof that the adaptations were produced by evolution. Creationists have never objected to the idea of natural selection as a mechanism for eliminating the unfit, non-adapted organisms. As a matter of fact, creationists long before Darwin were advocating natural selection as a *conservation* principle. The creationist Edward Blyth wrote on the subject at least 24 years before Darwin, and Loren Eiseley, a prominent modern evolutionist, has asserted that Darwin got the whole idea from Blyth. However, evolutionist Stephen Jay Gould has pointed out that many earlier creationists held similar views.

> Darwinians cannot simply claim that natural selection operates since everyone, including Paley and the natural theologians, advocated selection as a device for removing unfit individuals at both extremes and preserving, intact and forever, the created type.[1]

> Failure to recognize that all creationists accepted selection in this negative role led Eiseley to conclude falsely that Darwin has "borrowed" the principle of natural selection from his predecessor E. Blyth. The Reverend William Paley's classic work *Natural Theology,* published in 1803, also contains many references to selective elimination.[2]

---

1.   Stephen Jay Gould, "Darwinism and the Expansion of Evolutionary Theory," *Science* Vol. 216, April 23, 1982, p. 380.
2.   *Ibid,* p. 386.

As a screening device for eliminating the unfit, natural selection is a valid concept and, in fact, represents the Creator's plan for preventing harmful mutations from affecting and even destroying the entire species. And that is *all* it does! Yet evolutionists, especially Darwinians and neo-Darwinians have long insisted that it somehow "creates" new, better-adapted, more-fit species. In fact, Charles Darwin actually entitled his book *The Origin of Species by Natural Selection.*

Creationists have often pointed out, however, that the one concept Darwin's book did *not* discuss was the origin of species by natural selection! Darwin's evidences had to do with *varieties,* not *species,* and all else was conjecture. As the leading British evolutionist, Colin Patterson, has recently pointed out:

> No one has ever produced a species by mechanisms of natural selection. No one has ever gotten near it and most of the current argument in neo-Darwinism is about this question . . . .[1]

This "current argument in neo-Darwinism," interestingly enough, is merely echoing the arguments against neo-Darwinism that have been advanced by creationists for many years. Evolutionist Gould says:

---

1. Colin Patterson, "Cladistics," interview on British Broadcasting Corporation television program on March 4, 1982; producer Brian Leek, interviewer Peter Franz. Patterson is senior paleontologist at the British Museum of Natural History.

> The essence of Darwinism lies in a single phrase: natural selection is the creative force of evolutionary change. No one denies that selection will play a negative role in eliminating the unfit. Darwinian theories require that it create the fit as well.[1]

Herein, of course, we encounter the impotence of natural selection not only to produce the fit, but even to *define* the fit! Any definition is bound to be tautologous.

Now it is remarkable that, in this past seven years (1975-1982), practically all biologists have come to acknowledge this fact of redundancy. Natural selection is a force which somehow causes the survivors to survive. It enables those who adapt to adapt. Those who leave the most surviving descendants are the fittest to leave surviving descendants.

> One of the most frequent objections against the theory of natural selection is that it is a sophisticated tautology. Most evolutionary biologists seem unconcerned about the charge and only make a token effort to explain the tautology away. The remainder, such as Professors Waddington and Simpson, will simply concede the fact. For them, natural selection is a tautology which states a heretofore unrecognized relation: The fittest—defined as those who

---

1. Stephen Jay Gould, "The Return of Hopeful Monsters," *Natural History*, Vol. LXXXVI, June/July 1977, p. 28.

will leave the most offspring—will leave
the most offspring.

What is most unsettling is that some
evolutionary biologists have no qualms
about proposing tautologies as
explanations. One would immediately
reject any lexicographer who tried to
define a word by the same word, or a
thinker who merely restated his
proposition, or any other instance of gross
redundancy; yet no one seems scandalized
that men of science should be satisfied
with a major principle which is no more
than a tautology.[1]

The brilliant writer and vitalist philosopher,
Arthur Koestler, has incisively described the
quandry of the evolutionists—now widely
acknowledged, but still mostly ignored.

Once upon a time, it all looked so
simple. Nature rewarded the fit with the
carrot of survival and punished the unfit
with the stick of extinction. The trouble
only started when it came to defining
fitness . . . . Thus natural selection looks
after the survival and reproduction of the
fittest, and the fittest are those which
have the highest rate of reproduction, ...
we are caught in a circular argument
which completely begs the question of

---

1.   Gregory Alan Pesely, "The Epistemological Status of
     Natural Selection," *Laval Theologique et Philosophique*,
     Vol. XXXVIII, February 1982, p. 74.

what makes evolution evolve.[1]

Evolutionary literature is filled with marvelous stories of how organisms came to be so well adapted to their environments. These "just-so stories," these fairy tales for intellectuals, are, of course, pure imagination. No evolutionist can predict the course of future evolution, but he delights in "retrodicting" the wonders of past evolution. The reader of such tales should always take them *cum grano salis.*

> Paleontologists (and evolutionary biologists in general) are famous for their facility in devising plausible stories; but they often forget that plausible stories need not be true.[2]
> . . . All one can learn about the history of life is learned from systematics, from groupings one finds in nature. The rest of it is story-telling of one sort or another. We have access to the tips of a tree; the tree itself is theory and people who pretend to know about the tree and to describe what went on with it, how the branches came off and the twigs came off are, I think, telling stories.[3]

Nevertheless, as Koestler points out, even

---

1. Arthur Koestler, *Janus: A Summing Up.* (New York: Vintage Books, 1978), p. 170.
2. S. J. Gould, D. M. Raup, J. Sepkoski, L. J. M. Schopf, and D. S. Simberloff, "The Shape of Evolution: A Comparison of Real and Random Clades," *Paleobiology,* Vol. 3 (1), 1977, p. 34.
3. Colin Patterson, *op cit.* Patterson is a leading exponent of the new science of cladistics.

though the whole fabric of Darwinian and neo-Darwinian natural selection, as an explanation of evolution, is today in shreds, most intellectuals continue to hold it as an article of faith.

> In the meantime, the educated public continues to believe that Darwin has provided all the relevant answers by the magic formula of random mutations plus natural selection—quite unaware of the fact that random mutations turned out to be irrelevant and natural selection a tautology.[1]

In the meantime, if one is interested in reading a thorough critique, sound both scientifically and epistemologically, of natural selection as a causative and explanatory factor in evolution, the analysis of Professor R. H. Brady, of Ramapo College, is recommended.[2] Norman Macbeth, the Harvard lawyer whose 1971 book *Darwin Retried* was itself (even though not supporting creationism) a devastating critique of neo-Darwinism, says that one of Brady's papers "seemed to me to utterly destroy the entire idea of natural selection as presently conceived."[3]

---

1. Arthur Koestler, *op cit,* p. 185.
2. Ronald H. Brady, "Dogma and Doubt," *Biological Journal of the Linnaean Society,* Vol. 17, February 1982, pp. 79-96.
3. Norman Macbeth, "Darwinism: A Time for Funerals," interview in *Towards,* Vol. 2. Spring 1982, p. 18. The title has reference to the author's conviction that Darwinism and neo-Darwinism will die as soon as their older advocates die.

MUTATIONS AND MIRACLES

For at least the previous half-century, evolutionists had been supremely confident. The "evolutionary synthesis" promulgated by such leaders as Sir Julian Huxley, George Gaylord Simpson, Theodosius Dobzhansky, Ernst Mayr, J. B. S. Haldane, G. Ledyard Stebbins, Sewall Wright, Glen L. Jepsen, and others of similar stature, seemed to have solved all the major problems. The fossil record (allegedly) proved the *fact* of evolution, while mutations and natural selection (supposedly) provided the *mechanism.*

Creationists, however, kept on insisting that the fossil record showed no intermediate evolving forms, that natural selection was impotent and tautologous, and that mutations were all either neutral, harmful, or lethal, so that evolution still was based on no real evidence, either in the past or present. These arguments were all simply ignored.

But suddenly, all this has changed. A new wave of evolutionists has appeared, and these people are now using all the old creationist arguments themselves—not questioning evolution, of course (*that* is sacrosanct!)—but seeking to change the standard evidences (the fossil sequences have given way to protein sequences) and the Darwinian mechanisms (mutations are superseded by hopeful monsters and natural selection by chance).

The neo-Darwinians are not giving up easily, and the waters of the evolutionary sea have become very turbulent.

The neo-Darwinian theory of evolution is not only suffering from an identity crisis but may also be radically transformed to account for the growing number of scientific anomalies that continue to plague it. These were the underlying themes that could be inferred from presentations made by prominent scientists in the recently completed symposium entitled "What Happened to Darwinism between the Two Darwin Centennials, 1958-1982?" The symposium was convened under the auspices of the 148th Annual Meeting of the prestigious American Association for Advancement of Science held from January 3, 1982, to January 8, 1982, at Washington, D.C.[1]

There have been many similar symposia held in recent years, as well as many articles both criticizing and defending neo-Darwinism in various periodicals. The conflict occasionally has become quite heated, though evolutionists of all points of view are quite unified when it comes to fighting creationism.

The symposium was a disappointment to the true believers of neo-Darwinism. Implicit in their counter-offensive to stamp out creationism was the recognition that they had to contain and mend the fissures that were increasingly undermining the scientific foundation of their own neo-Darwinist position. To their dismay, the

---

1. Nicky Perlas, "Neo-Darwinism Challenged at AAAS Annual Meeting," *Towards*, Vol. 2, Spring 1982, p. 29.

Provine symposium aggravated and
deepened the fissures.[1]

Neo-Darwinism, of course, had concluded
that random mutations in the genetic systems
of organisms provided the basic materials on
which natural selection could act to produce
new, better-equipped species. The problem,
repeatedly stressed by creationists, was that
mutations are neutral or harmful, not helpful.
At least that is true of all known mutations.
Evolutionary theory regarding past mutations
tends to ignore or denigrate this fact, but it is
always emphasized when dealing with
environmental hazards which might cause
present-day mutations. Several years ago, the
Environmental Mutagenic Society made a
detailed study of this subject and concurred
that mutations should always be avoided if
possible.

Most mutations producing effects large
enough to be observed are deleterious,....
Furthermore, the wide variety of
mechanisms by which radiations and
chemicals induce mutations make it very
unlikely that generalized schemes can be
devised to protect against mutagens,
except by avoiding them in the first place.[2]

The exact nature of gene mutations is still
somewhat obscure, since the exact nature of

---

1. Nicky Perlas, *op cit,* p. 30.
2. Environmental Mutagenic Society, "Environmental
    Mutagenic Hazards," *Science,* Vol. 187, February 14,
    1975, p. 503.

genes is still uncertain. Whatever it is, however, a mutation represents an unpredictable—evidently random—change in an extremely complex genetic programmed system. Since the program is thus inadvertently changed, it represents a *mistake* in the transmission of genetic information.

> Being an error process, mutation consists of all possible changes in the genetic material (excluding recombination and segregation).[1]

The Society therefore recommends—as do practically all other scientists—that everything feasible be done to eliminate radiations and mutagenic chemicals from the environment.

> Since the vast majority of detectable mutations are deleterious, an artificially increased human mutation rate would be expected to be harmful in proportion to the increase.[2]

Evolutionists are always careful to say only that the "vast majority" of mutations are harmful, leaving open the possibility that some just might be beneficial. The possibilities are very limited, however.

> From the standpoint of population genetics, positive Darwinian selection represents a process whereby advantageous mutants spread through the species. Considering their great

1.   Environmental Mutagenic Society, *op cit*, p. 504.
2.   *Ibid*, p. 512.

importance in evolution, it is perhaps
surprising that well-established cases are
so scarce; for example, industrial
melanisms in moths and increases of DDT
resistance in insects are constantly being
cited. [1]

As a matter of fact, neither of the cases cited is
a true mutation. Industrial melanism in moths
is simply a recombination of genetic factors
already present, and the same is true of the
insects.

Insect resistance to a pesticide was first
reported in 1947 for the housefly (Musca
domestica) with respect to DDT. Since then
the resistance to pesticides has been
reported in at least 225 species of insects
and other arthropods. The genetic variants
required for resistance to the most diverse
kinds of pesticides were apparently
present in every one of the populations
exposed to these man-made compounds.[2]

Because of all these problems, more and
more evolutionists are realizing that ordinary
mutations, as actually observed in nature, are
not at all adequate to provide the basis of
evolution. Neither will it do to suppose that
very small, non-observable, mutations,
gradually accumulating over long periods of
time, could do the job.

---

1.   Motoo, Kimura, "Population Genetics and Molecular
     Evolution," *Johns Hopkins Medical Journal,* Vol. 138,
     June 1976, p. 260.
2.   Francisco Ayala, "The Mechanisms of Evolution,"
     Scientific American, Vol. 239, September 1978, p. 63.

Bacteria, the study of which has formed a great part of the foundation of genetics and molecular biology, are the organisms which, because of their huge numbers, produce the most mutants . . . . bacteria, despite their great production of intra-specific varieties, exhibit a great fidelity to their species. The bacillus *Echerichia coli,* whose mutants have been studied very carefully, is the best example. The reader will agree that it is surprising, to say the least, to want to prove evolution and to discover its mechanisms and then to choose as a material for this study a being which practically stabilized a billion years ago.[1]

The same is true of the organism whose mutants have probably been studied more than any other.

The fruit-fly *(Drosophila melanogaster),* the favorite pet insect of the geneticists, whose geographical, biotopical, urban, and rural genotypes are now known inside out, seems not to have changed since the remotest times.[2]

The author of the above evaluations, Pierre Grassé is not a creationist and, in fact, as France's leading zoologist, held the Chair of Evolution at the Sorbonne (France's leading university) for 20 years. His opinion of

---

1. Pierre P. Grassé, *Evolution of Living Organisms (New York: Academic Press, 1977), p. 87.*
2. *Ibid,* p. 130.

mutations, as an explanatory cause of evolution, is summarized below:

> The opportune appearance of mutations permitting animals and plants to meet their needs seems hard to believe. Yet the Darwinian theory is even more demanding: a single plant, a single animal would require thousands and thousands of lucky, appropriate events. Thus, miracles would become the rule: events with an infinitesimal probability could not fail to occur . . . . There is no law against day-dreaming, but science must not indulge in it.[1]

This is surely an insightful evaluation, and Grassé's fellow evolutionists would do well to pay attention to it. Mutations and natural selection must have been energized by a continuous succession of miracles if they really do constitute the explanation of evolution. It is small wonder that the new school of evolutionists is searching for a better explanation.

CHAOTIC GENETICS

Since the old ideas of small, random point mutations have proven inadequate to explain evolution—and genetic mutations are now needed which will produce hopeful monsters or other large changes—the field of genetics has proliferated in all sorts of strange directions. Instead of the old Mendelian concept of a specific gene for each physical characteristic,

---

1.   Grassé, *op cit,* p. 103.

modern molecular biology has introduced an amazing assortment of genes, and their corresponding DNA, into the literature. There are structural genes, regulatory genes, selfish genes, coding genes, junk genes, split genes, pseudogenes, processed genes, jumping genes, repeating genes, satellite genes, converted genes, and various others. Each functional gene affects many characteristics; every characteristic is affected by many genes, but most genes seem to do nothing at all. Many gene sequences repeat themselves in various organisms, and in most cases the function of these repeating sequences is quite unknown. Then there are the chromosomes, the DNA molecules, the RNA, the cortex, and various other components of the cell, all playing various key roles in heredity, but all still very imperfectly understood.

This is hardly an appropriate place to try to explain all these terms and concepts. Even specialists in molecular biology are still trying to sort it out. A little seems to be known about many things, but not much is known about anything specific in this unique field of study.

> The all-pervading message of the Cambridge meeting was that genomic DNA is a surprisingly dynamic state . . . . The most obvious comment to make about the genomes of higher organisms is that biologists understand the function of only a tiny proportion of the DNA in them . . . .[1]

---

1. Roger Lewin, "Do Jumping Genes Make Evolutionary Leaps?" *Science*, Vol. 213, August 7, 1981, p. 634.

Obviously it is hoped that this mysterious world of molecular biology will eventually provide desperately needed answers (needed by evolutionists, that is) as to what causes evolution. It must be embarrassing to be certain evolution is true and yet to have no idea how it works!

Plenty of money, over three billion dollars per year, is being devoted to molecular biology, so evolutionists are hopeful that an explanation will one day be forthcoming.

But it hasn't yet! Creationists are amazed at the strong faith exercised by evolutionists. Over a century of intensive research into mechanisms of biologic change has still failed to turn up a plausible genetic model to explain evolution. Evolution is supposed to be an all-pervading process, still actively going on, yet no one has ever seen it happen, there are no evolutionary transitions recorded in the fossils, and no known genetic mechanism is capable of producing it. Yet evolutionists repeatedly insist that evolution is a proven fact of science! This is an amazing commentary on human nature in its perverse attempt to get rid of God.

HEMOGLOBIN HIERARCHIES

However, genetic studies have provided a new set of data which evolutionists have started using again in recent years as an evidence for evolution. With the increasing recognition that the fossil record, as well as the mutation/selection mechanism of neo-Darwinism, really constitutes an argument against evolution instead of for evolution, a new line of evidence was urgently needed. Apparently the best they could come up with is the supposed hierarchical arrays of various proteins in different organisms, which are supposed to correspond to the chronological histories of their respective times of divergence from the ancestral line of common ancestry. This is the so-called evidence from molecular homology.

For example, it is commonly asserted now that man and the chimpanzee must be very closely related because they are said to share 99% of their DNA. Similarly the chimpanzee is allegedly man's closest relative based on hemoglobin similarities. By such comparisons it is asserted that a complete evolutionary family tree can be derived between man and all other organisms, with the time of divergence of each from the ancestral stock actually constituting a "molecular clock," which can even be used to give absolute dates by calibrating against radiometric ages. It is even alleged that the same evolutionary relationships are obtained from most other proteins in the various organisms and that these all correspond to the paleontological record as well.

Now even if this were all true, it certainly would not prove evolution. This is nothing more or less than the old argument from similarities, or comparative morphology. One of the traditional arguments for evolution has always been that of similarities—similarities in anatomy, similarities in embryology, etc. This, of course, proves nothing. Similarities indicate a common Designer at least as much as they indicate a common ancestor. Why should it be surprising that chimpanzee DNA should be very similar to human DNA? The entire structure of chimpanzees is far more similar to that of humans than is, for example, that of fishes or scorpions.

The DNA, for that matter, even though it seems to carry the genetic information for all organisms, is mostly of uncertain function.

> In the human genome, for instance, these protein-coding genes constitute marginally more than one percent of all the DNA. The rest of the genome is the target of much speculation, but few secure answers.[1]

It would seem that, if the function of 99% of human DNA is unknown, the fact that 99% of the 1% of human DNA which *is* known corresponds to chimpanzee DNA really proves very little. It is only the one percent difference, apparently, that does all the coding (if indeed, the DNA is really responsible for all such genetic coding, a proposition which has never been satisfactorily demonstrated).

Furthermore, the supposed similarities have been much over-rated. Blood proteins and Cytochrome c are two types of proteins which have often been cited as supporting the imaginary evolutionary hierarchy. Yet note the following evaluations of these two systems by two eminent biologists.

> If blood proteins are a representative sample of proteins coded by structural

---

1.   Lewin, *op cit*, p. 634.

genes, the most similar species should have the most similar blood proteins. Wilson (Univ. Calif. of Berkeley) and his colleagues found, however, that structural genes for blood proteins accumulate mutations at rates that appear independent of anatomical evolution . . . . It thus seems evident that the old method of comparing proteins of different species may no longer be the primary tool for investigating the mechanisms underlying the evolution of organisms.[1]

The cytochrome c of man differs by 14 amino acids from that of the horse, and by only 8 from that of the kangaroo. Similar facts are found in the case of hemoglobin; the chain of this protein in man differs from that of the lemurs by 20 amino acids, by only 14 from that of the pig, and by only 1 from that of the gorilla. The situation is practically the same for other proteins.[2]

By all other physical measures, of course, man should have been much more closely related to the horse than the kangaroo and the lemur than the pig. In a very provocative paper presented at the American Museum of Natural History, Dr. Colin Patterson, senior paleontologist at the British Museum of Natural History, cited several such anomalies and

1. Gina Bari Kojata, "Evolution of DNA: Changes in Gene Regulation," *Science*, Vol. 189, August 8, 1975, pp. 446-447.
2. Pierre P. Grassé, *Evolution of Living Organisms* (New York: Academic Press, 1977), p. 194.

contradictions in the molecular data. Even though he is a leading evolutionist, he then concluded:

> In other words, evolution may very well be true, but basing one's systematics on it will give bad systematics.[1]

Systematics, of course, is the discipline that tries to classify organisms in appropriate taxonomic groups supposed to reflect relationships.

There are even more difficulties that become apparent when these supposed molecular homologies are used as a molecular "clock." A recent authoritative study of this subject first defines this so-called clock as follows:

> The fundamental tenet of the molecular clock hypothesis is that evolutionary rates of homologous proteins are regular, so that the interval separating living species from common ancestors is reflected in the degree of protein dissimilarity between them.[2]

This assumption of uniform protein divergence rates, that is, of constant nucleotide mutation rates, is an assumption implicitly

---

1. Colin Patterson, "Evolutionism and Creationism." Speech at American Museum of Natural History, New York, N.Y., Nov. 5, 1981, p. 14 of transcript.
2. Kenneth A. Korey, "Species Number, Generation Length, and the Molecular Clock," *Evolution*, Vol. 35, No. 1, 1981, p. 139.

based on neo-Darwinian population genetics. Korey shows that the critical factors of generation length and numbers of individuals in the population have been ignored. This invalidates the whole procedure.

> Coupled with complementary findings regarding the significance of species bottlenecks to protein divergence rates, these effects undermine the main premise of the clock thesis, especially as it applies to the dating of lineages not remotely separated. [1]

Steven Stanley, who has rejected such gradualistic and uniformitarian premises for other reasons, is much less enamored with his molecular homology idea than are the more traditional evolutionists.

> Simple estimates of overall genetic distance between species reveal little about degrees and rates of morphologic divergence.[2]

Korey has pointed out that the method as currently used will regularly give ages that are too "short" (that is, will underestimate the difference between the two organisms being compared, and thus will underestimate the time since their assumed evolutionary divergence from a common ancestor). In

---

1. *Ibid,* p. 146.
2. Steven M. Stanley, *Macroevolution: Pattern and Process* (San Francisco: W. H. Freeman and Co., 1979), p. 61.

particular he criticizes the supposed close similarity between man and chimpanzee which has been inferred from the uniformitarian premise of the molecular clock.

> Certainly the widely contested date for the separation of *pan* and *Homo* that Sarich and Wilson suggest is subject to this bias.[1]

Drs. Vincent Sarich and A. C. Wilson, of the University of California at Berkeley, are the two scientists primarily responsible for this current widespread notion that man, chimpanzee, and gorilla diverged from their "common ancestor" only about five million years ago, as based on these interpretations of molecular chronologic homology. Presumably, as the still more current notions of fluid genomes, jumping genes, and chaotic genetics take over, the uniformitarianism of the molecular clock interlude in evolutionary thinking will quietly disappear.

In a review of a recent symposium on the applications of these molecular homologies to evolutionary anthropology, the reviewer makes the following cynical, but appropriate concluding comment:

> On the current state of theoretical evolutionary work as described in this volume, I quote higher authority: the Red King acting as judge in *Alice in*

---

1.   Kenneth A. Korey, *op cit*, p. 145.

Wonderland. . . . "If there's no meaning in it, that saves a lot of trouble, you know, as we needn't try to find any."[1]

---

1.   Roy J. Britten, Review of *Molecular Anthropology,* Ed. by Morris Goodman, R. E. Tashian, and J. H. Tashian (New York: Plenum, 1976, 466 pp.), in *Science,* Vol. 198, October 21, 1977, p. 287.

Chapter 3

Where Have All The Fossils Gone?

THE CASE OF THE LOST CONNECTIONS

For a long time, the fossil record was cited as the main proof of evolution. The actual, documented, history of the evolution of life during the earth's geological ages was, so it was said, preserved for all to see, in the fossils in the sedimentary rocks of the crust of the earth. Even though evolution proceeded too slowly to be observed in action, the fossil record proved it had occurred through the past. The mechanism of evolution might be uncertain, but the *fact* of evolution was established by the fossils.

But this also has changed in the past seven years! Dr. Mark Ridley, of Oxford University's Department of Zoology, now says:

> In any case, no real evolutionist, whether gradualist or punctuationist, uses the fossil record as evidence in favor of

the theory of evolution as opposed to
special creation . . . . [1]

This is really quite an amazing flip-flop, and
such a statement would have been unthinkable
a decade ago. The fossil record had always
been an integral component of neo-Darwinism,
fully supporting (so they claimed) the concept
of slow and gradual evolution over a billion
years of earth history.

However, for many years now, creationist
scientists have been abundantly documenting
—both in public lectures and in written
expositions—the fact that the fossil record is
completely devoid of any transitional forms.
Suddenly, almost overnight it seems,
paleontologists and assorted other evolutionary
scientists, are saying the same thing. They are
using the same arguments and evidences
creationists have been using for years, not
against evolution in general, but against slow-
and-gradual evolution.

At the higher evolutionary transition
between basic morphological designs,
gradualism has always been in trouble,
though it remains the "official" position of
most Western evolutionists. Smooth
intermediates between *Baupläne* are
almost impossible to construct, even in
thought experiments. There is certainly no
evidence for them in the fossil record

----

1.   Mark Ridley, "Who Doubts Evolution?" *New Scientist*,
       Vol. 90, June 25, 1981, p. 831.

(curious mosaics like *Archaeopteryx* do not count).[1]

The authors of the above statement, Niles Eldredge and Stephen Jay Gould, are widely esteemed these days as the founding fathers of the "punctuated equilibrium" school of evolutionary thought, which offers saltationism (evolution by jumps) in place of gradualism (slow-and-gradual evolution).

Note especially their admission that the fossil record (with which they probably are as familiar as any living paleontologists) does not offer any intermediate forms between major morphologic types. They call these types *Baupläne* (a German word meaning "building plans," basic designs), a term which apparently corresponds nicely with what creationists call "the Genesis *kinds.*" There are many transitional forms *within* kinds, but none *between* them.

Note also the important acknowledgment that *Archaeopteryx* does not really "count" as an intermediate form, since it is merely a "curious mosaic." This is exactly what creationists have always maintained, even while evolutionists have been proudly parading *Archaeopteryx* as their best example of a major transitional form.

Herein is a very important distinction, of

---

1. Stephen J. Gould and Niles Eldredge, "Punctuated Equilibria: The Tempo and Mode of Evolution Reconsidered," *Paleobiology*, Vol. 3, Spring 1977, p. 147.

course. A true transitional form should have transitional structures. That is, a reptile in the process of becoming a bird should have "sceathers" (half-scales, half-feathers) and "lings" (half-legs, half-wings), but *Archaeopteryx* had fully developed, aerodynamically efficient feathers and fully functional wings.[1] Its so-called "reptilian" characteristics (e.g., teeth) were also fully functional, not atrophied or vestigial. In other words, *Archaeopteryx* was a "mosaic" of fully functional structures, some of which modern taxonomists choose to call reptilian, some of which they identify as avian. It was *not* a "transition," with inefficient changing structures. Furthermore, it is now well known that fossils of true birds have been found in strata as old or older than those associated with *Archaeopteryx*.[2] Whatever the latter may be, it cannot be the ancestor of the birds, because they are at least as old as *Archaeopteryx!*

It should be understood that the created "kinds" (or *Baupläne*, if evolutionists prefer a non-Biblical term) have no necessary correlation with the arbitrary categories of modern taxonomy. If an animal is properly

1.   Storrs L. Olson and Alan Feduccia, "Flight Capability and the Pectoral Girdle of *Archaeopteryx*," *Nature*, Vol. 278, March 15, 1979, pp. 247-248.
2.   Jean L. Marx, "The Oldest Fossil Bird: A Rival for *Archaeopteryx?*" *Science*, Vol. 199, January 20, 1978, p. 284.

designed for its environment, with no clearly transitional structures, it should be regarded as a distinct kind, regardless of whether it fits a particular "order" or "class." All such taxonomic groupings have been defined arbitrarily in accordance with human judgment, not by divine revelation.

The fact is, that there are *no* true transitional forms (in the sense of transitional or incipient structures) that have ever been documented out of the billions of fossils known to be preserved in the rocks of the geologic column. This is hardly surprising, since such intermediate structures would be useless in the struggle for existence. An animal with a half-leg/half-wing, for example, could neither run nor fly. Gould has noted this as follows:

> The absence of fossil evidence for intermediary stages between major transitions in organic design, indeed our inability, even in our imagination, to construct functional intermediates in many cases, has been a persistent and nagging problem for gradualistic accounts of evolution.[1]

That is, not only have intermediate forms not been found, they cannot even be imagined!

This is exactly what creationists have been stressing for years, but evolutionists have been

---

1. Stephen Jay Gould, "Is a New and General Theory of Evolution Emerging?" *Paleobiology*, Vol. 6, January 1980, p. 127.

maintaining otherwise until recently. Suddenly there has developed a wide acknowledgment that the creationists were right all along (evolutionists don't put it *that* way, of course!). Dr. D. V. Ager, then President of the British Geological Association, in his 1976 presidential address, said, for example:

> It must be significant that nearly all the evolutionary stories I learned as a student . . . have now been debunked.[1]
> The point emerges that, if we examine the fossil record in detail, whether at the level of orders or of species, we find—over and over again—not gradual evolution, but the sudden explosion of one group at the expense of another.[2]

In spite of such admissions, which are frequently made by the "punctuated equilibri-umites" when they are opposing the "neo-Darwinites," both types of evolutionists *still* claim that there *are* transitional forms when either group fights the creationists. However, the transitional forms cited are almost always either *Archaeopteryx* or the so-called mammal-like reptiles (both of which are simply "mosaics" of fully functional structures) or

---

1. D. W. Ager, "The Nature of the Fossil Record," *Proceedings of the Geological Association*, Vol. 87, No. 2, 1976, p. 132.
2. *Ibid*, p. 133.

such series as those of the horses or the hominids.

But none of these (and the same is true of other less-publicized supposed transitional fossils) qualify as true evolutionary transitions or intermediates. They fail on *three* counts: (1) none of their features are really *transitional,* but all fully functional, optimally designed for their respective environments; (2) each one is quite distinct from all others in the group, with no *gradual* changes documented; (3) their chronologies are not properly in evolutionary order, but are overlapping and contradictory, representing a "bush," not a "tree."

An additional word is needed about the mammal-like reptiles, since many evolutionists have taken to citing these as examples of transitional forms. As just noted, these creatures should be regarded as created kinds, with all their structural features perfectly equipped to function in their respective environments. There were hundreds of species of these creatures, and they supposedly arose about the same time in evolutionary history as the other reptiles. In fact they became extinct even before the age of the dinosaurs. Exactly which reptile evolved into the first mammal-like reptile and which mammal-like reptile evolved into a mammal is entirely unknown:

> Each species of mammal-like reptile that has been found appears suddenly in the fossil record and is not preceded by the species that is directly ancestral to it. It disappears some time later, equally abruptly, without leaving a directly

descended species, ...[1]
   The transition to the first mammal,
which probably happened in just one or, at
most, two lineages, is still an enigma.[2]

It is obviously only the urgent necessity to
defend evolution that justifies calling such
animals transitional forms. Both the reptiles
and the mammals survived nicely through all
the ages, but all these supposed evolutionary
intermediaries between them died out even
before the so-called age of reptiles began.

   Furthermore, since both mammals and
reptiles are land vertebrates, their skeletal
structures are necessarily very similar. It is not
primarily their *bones* which distinguish
mammals from reptiles; it is their reproductive
systems and other soft parts that leave no
trace in the fossils. The only bones that are
different are certain small bones associated
with the middle ear and the masticatory
apparatus. The mammal-like reptiles are
supposed to have had a combination of
reptilian and mammalian features in these
particular bones, but this in no way proves any
kind of evolutionary connection. The term
"mosaic" is again far more appropriate and
realistic.

1.   Tom Kemp, "The Reptiles that Became Mammals," *New
Scientist,* Vol. 92, March 4, 1982, p. 583. Dr. Kemp is
Curator of Zoological Collections at the Oxford
University Museum in England.

2.   Roger Lewin, "Bones of Mammals' Ancestors Fleshed
Out," *Science,* Vol. 212, June 26, 1981, p. 1492.

A recent book[1] is supposed to be an authoritative compendium of all the data on a major aspect of this subject. However, a reviewer well qualified to evaluate the book and this whole question says:

> These general statements about the evolution of the mammalian middle ear that appear are in the nature of proclamations. No methods are described which allow the reader to arrive with Fleischer at his "ancestral" middle ear, nor is the basis for the transformation series illustrated for the middle ear bones explained . . . . Those searching for specific information useful in constructing phylogenies of mammalian taxa will be disappointed.[2]

It is obvious without going into the technicalities of the morphology and stratigraphy of these long-extinct animals that they cannot really be any kind of evolutionary connecting link between reptiles and mammals.

The famous evolutionary family tree of the horse (that is, *Eohippus* to *Equus,* throughout the Tertiary Period) fares even worse. The first in the series, *Eohippus,* was probably a hyrax rather than a horse, and should not even be included in the group at all. He is now again

---

1. Gerald Fleischer, *Evolutionary Principles of the Mammalian Middle Ear* (Berlin: Springer-Verlag, 1978).
2. R. Eric Lombard, "Review of *Evolutionary Principles of the Mammalian Middle Ear," Evolution,* Vol. 33, No. 4, 1980, p. 1230.

being called *Hyracotherium* (which was the name given by its first discoverers, before certain North American paleontologists got the quaint notion that it might be the ancestor of the horses).

Furthermore, this animal was quite stable, in terms of the standard geological column. Paleontologist Stephen Stanley notes:

> Those who in the past have contemplated the formation of the modern horse by gradual evolution, beginning with this early genus, must now contend with the fact that at least two species of *Hyracotherium* lasted for several million years without appreciable change.[1]

As a matter of fact, there has been much less change than evolutionists contemplate, since *Hyracotherium* is clearly much the same as our modern hyrax. With respect to the modern horse, *Equus*, Stanley notes that it also has been highly stable.

> It is notable that the evidence of great stability for species of *Hyracotherium* is complemented at the other end of equid phylogeny, by data showing that ten species of horses lived through most or all of Pleistocene time . . . .[2]

As far as the intermediate horses are concerned (*Miohippus, Merychippus,*

---

1.   Stephen M. Stanley, "Macroevolution and the Fossil Record," *Evolution*, Vol. 36, No. 3, 1982, p. 464.
2.   *Ibid.*

Pliohippus, etc.), it is now well known that they no longer diagram into a tree, but a "bush." There is much overlapping and each is separated from the others without intermediate forms between. In the John Day strata of Oregon, the three-toed grazer *Merychippus* has been found in the same formation with the single-hoofed *Pliohippus,* for example. No wonder David Raup says:

> The record of evolution is still surprisingly jerky and, ironically, we have even fewer examples of evolutionary transition than we had in Darwin's time. By this I mean that some of the classic cases of Darwinian change in the fossil record, such as the evolution of the horse in North America, have had to be discarded or modified as a result of more detailed information . . . . [1]

Now if *Archaeopteryx* and the horses and the mammal-like reptiles do not constitute transitional forms (and, as we have shown, they do not), and if these are the ones most cited by evolutionists as examples of transitional forms (and they are), then we are abundantly justified in concluding there are no real transitional forms anywhere at all in the fossil record. This is amazing, if evolution is true. Statistically there should be many transitional forms. In

---

1. David M. Raup, "Conflicts between Darwin and Paleontology," *Field Museum of Natural History Bulletin,* Vol. 50, January 1979, p. 25. Raup is Curator of Geology at the Field Museum.

fact, if evolution is really true, it would seem
logical that *all* creatures, living and extinct,
should be transitional forms, in the process of
evolving into new, higher forms. And yet there
are none!

MEN, MONKEYS, AND MISSING LINKS

Of all the supposed evolutionary family trees,
the one leading to man should be best
documented. As the most recent evolutionary
arrival, pre-human fossils have been exposed to
decay processes for the shortest period of time,
and so should be better preserved and easier to
find than any others. Furthermore, since they
are of greatest interest to man, more people
have been looking for them than for any other
kinds of fossils. Consequently, if there are any
real transitional forms anywhere in the fossil
record, they should be most abundantly
documented in the line leading from the first
primate to modern man. Certainly the finds in
this field have been more publicized than in
any other.

And yet after over 100 years of intensive
searching, none have been found. The links are
still missing. A few "hominid" bones and teeth
have been found which some anthropologists
(not all, by any means) have argued were in the
line of human evolution, but all of these are
very doubtful, highly controversial, and
extremely fragmentary. The general public has
no idea just how fragmentary!

> The fossils that decorate our family tree
> are so scarce that there are still more
> scientists than specimens. The remarkable
> fact is that all the physical evidence we

have for human evolution can still be placed, with room to spare, inside a single coffin.[1]

And the obvious comment that one is almost compelled to make is that a coffin is exactly *where* it should be placed.

One of the nation's leading paleoanthropologists, Yale's David Pilbeam, makes a similar comment about the scarcity of data:

> Human paleontology shares a peculiar trait with such disparate subjects as theology and extraterrestrial biology; it contains more practitioners than objects for study.[2]

There are not very many practitioners either, for that matter. Physical anthropology has had an impact on human thought far outweighing the normal influence of a very small and otherwise insignificant discipline. More recently, Pilbeam has admitted the following:

> I know that at least in paleoanthropology, data are still so sparse that theory heavily influences interpretations. Theories have, in the past, clearly reflected our current ideologies instead of the actual data.[3]

---

1. Lyall Watson, "The Water People," *Science Digest,* Vol. 90, May 1982, p. 44
2. David Pilbeam and Stephen J. Gould, "Size and Scaling in Human Evolution," *Science,* Vol. 186, Dec. 6, 1974, p. 892.
3. David Pilbeam, "Rearranging Our Family Tree," *Human Nature,* June 1978, p. 45.

Pilbeam, of course, is referring not to evolution in general, but to particular ideas about the sequences of human evolution. He is not about to become a creationist! Nevertheless, it seems odd that evolutionists can be so sure about evolution when they admittedly know nothing as to how it works or what course it has followed.

Actually, no more is known about the evolution of apes (pongids) than that of man.

> Unfortunately, the fossil record of pongids is nonexistent, making a glaring deficiency in the whole story.[1]

How is it, then, that evolutionists are so dogmatically insistent that apes and people have a common ancestor? So far as all real evidence goes, apes have always been apes and people have always been people. Yet such anthropologists as Carl Johanson and the Leakeys have become world famous because of their excavation of a few handfuls of bones in Africa which they claim give evidence of human evolution. Interestingly enough, Richard Leakey and Johanson disagree heatedly as to where their respective finds (known popularly as Lucy, Skull 1470, Handy Man, etc.) fit in the line of human evolution, but they are united in their opposition to any suggestion of human creation.

Some evolutionists (including evolutionist debaters) are so anxious to persuade people

---

1. *Ibid*, p. 43.

about man's evolution that they have deliberately ignored the physical evidence in order to arrange these so-called hominid fossils into what looks like an evolutionary family tree. They have, for example, taken all relevant data on cranial capacities, from the 500 c.c. capacities of the Australopithecines (and modern gorillas) to the 1500 c.c. skulls of modern man, and published them in a graphical sequence purporting to show the gradual increase in brain size as man evolved. However, this "evolutionary line" is not a "time line," for the dates assigned to these skull fragments are widely divergent, with considerable evidence that *Australopithecus, Homo erectus,* and *Homo sapiens* have all lived contemporaneously in the past. As a matter of fact, one could arrange the skull sizes of living human beings in a similar series, increasing from as low as 700 c.c. to well over 2000 c.c., and this would have nothing whatever to do with evolution, since cranial capacity has no necessary correlation at all with human intelligence or mental ability—still less with the soul/spirit complex which completely separates man from all animals.

Modern evolutionists, like the evolutionary pantheists of antiquity, may "not *like* to retain God in their knowledge" (Romans 1:28), but that does not justify them in distorting the overwhelming evidence that man was specially created in God's image and then trying to make it appear that he is merely another animal and that God Himself is merely an imaginary construct created in man's image. "Professing themselves to be wise, they became fools, and changed the glory of the uncorruptible God into an image made like to corruptible man, and to birds, and fourfooted beasts, and creeping things" (Romans 1:22-23).

The fact is that, even after a century of intensive searching and special pleading, there is still no real fossil evidence of human evolution.

> Not surprisingly, despite the diligent research done in East Africa by paleontologists Richard Leakey and Donald Johanson, there are gaping holes in the evolutionary record, some of them extending for four to six million years.
>
> Modern apes, for instance, seem to have sprung out of nowhere. They have no yesterday, no fossil record. And the true origin of modern humans—of upright, naked, toolmaking, big-brained beings—is, if we are to be honest with ourselves, an equally mysterious matter.[1]

Even the professional paleoanthropologists, though still firmly committed to an evolutionary faith, are in disarray as to the actual fossil histories.

> All this makes a much more complex picture of hominoid evolution than we once imagined. It no longer resembles a ladder but is, instead, more like a bush. . . . Hominids evolved, as did many other mammal groups, with diverse and overlapping radiations. There is no clear-

---

1.   Lyall Watson, *op cit,* p. 44.

cut and inexorable pathway from ape to human being.[1]

Anthropologists are like the blind men looking at the elephant, each sampling only a small part of the total reality.[2]

The simple idea of evolution, which it is no longer thought necessary to examine, spreads like a tent over all those ages that lead from primitivism into civilization. Gradually, we are told, step by step, men produced the arts and crafts, this and that, until they emerged in the light of history. Those soporific words "gradually" and "step-by-step," repeated incessantly, are aimed at covering an ignorance which is both vast and surprising. One should like to inquire: Which steps? But then one is lulled, overwhelmed, and stupefied by the gradualness of it all, which is at best a platitude, only good for pacifying the mind, since no one is willing to imagine that civilization appeared in a thunderclap.[3]

Any discussion of specific fossils is out of date almost as soon as published, so it is almost redundant to critique the hominids currently in vogue. The once-fashionable names of Java Man, Piltdown Man, Nebraska Man, Heidelberg Man, Rhodesia Man, Peking Man,

1. David Pilbeam, *op cit,* pp. 44, 45.
2. Alan Mann (anthropologist at University of Pennsylvania), as quoted in *Time* magazine article, "Puzzling Out Man's Ascent," November 7, 1977, p. 77.
3. Giorgio de Santillana and Hertha von Dechend, *Hamlet's Mill* (Boston: Gambit, Inc., 1969), p. 68.

and others which used to be offered as proof of man's evolution, are nowadays all but ignored in anthropological discussions. Neanderthal Man and Cro-Magnon Man are now universally accepted as *Homo sapiens* today. Even *Ramapithecus* is now out of favor as an early hominid.

Currently (that is, as of 1982) *Australopithecus* is receiving the most attention, and this group includes those fossils which once were assigned more specific roles (e.g., *Zinjanthropus, Homo habilis,* Skull 1470, Lucy). Although some scientists and many popular writers, believe man is descended from the australopithecines, many do not.

> Interestingly, despite almost a decade of technically sophisticated analyses of australopithecine remains, there is still considerable controversy over their functional and phylogenetic significance— in particular whether they are too divergently specialized to be considered suitable ancestors for *Homo*.[1]

The most sophisticated of these analyses were performed by Solly Zuckerman, Charles Oxnard, and their colleagues. These multivariate statistical analyses were computerized and highly detailed, showing almost conclusively that the australopithecines were some form of

---

1.   James A. Hopson and Leonard B. Radinsky, "Vertebrate Paleontology; New Approaches and New Insights," *Paleobiology,* Vol. 6, Summer 1980, p. 263.

extinct ape, and that they did not walk erect. Oxnard's evaluation was summarized as follows:

> Although most studies emphasize the similarity of the australopithecines to modern man, and suggest, therefore, that these creatures were bipedal tool-makers . . . a series of multivariate statistical studies of various postcranial fragments suggests other conclusions.[1]

The conclusion was, essentially, that *Australopithecus* was an extinct ape, more like the orangutan than any other living creature.

Some have argued, however, that Mary Leakey's 1978-79 finds of fossil footprints proved that *Australopithecus* did walk erect. These footprints indicated a bipedal creature, but the only reason for identifying them with *Australopithecus* was the fact that they were found in Africa and were dated radiometrically to correspond to the assumed age of *Australopithecus*. Consider, however, the following evaluation of them:

> The uneroded footprints show a total morphological pattern like that seen in modern humans . . . . Spatial relationships of the footprints are strikingly human in pattern . . . . The Laetoli hominid trails at site G do not differ substantially from

---

1.   Charles Oxnard, *Nature,* Vol. 258, 1975, p. 389.

modern human trails made on a similar
substrate.[1]

In other words, these trails are
indistinguishable from trails of true human
footprints. Why, then, try to make them out to
be australopithecine footprints? Why not draw
the much more reasonable conclusion either
that the dates are wrong or else that man lived
at the same time as the australopithecines?[2]

The latter conclusion, while contrary to the
usual evolutionary prejudices, is not really so
far-fetched, even in the framework of the
standard geologic ages. It is now rather
generally agreed by anthropologists that
australopithecines were contemporaries of
Homo erectus, even though some believe the
latter had evolved from the former. If that is
the case, why could not *Homo sapiens* have
been contemporaneous with both? Richard
Leakey, for example, has described localities
where fossils of *Homo erectus* and

---

1. P. H. Busse and K. E. Heikes, "Evolutionary Implications
    of Pliocene Hominid Footprints," *Science,* Vol. 208, April
    11, 1980, p. 175. These trails, in fact, look very much
    like the trails of human footprints found in association
    with dinosaur footprints in the Glen Rose Formation of
    central Texas.
2. University of Chicago anthropologist Russell Tuttle says
    that the Laetoli footprints are "virtually human," that
    they "do not match the foot bones found in Hadar," and
    that, in fact, Lucy's pelvis was "better suited for
    climbing than for walking." (W. Herbert, "Was Lucy a
    Climber? Dissenting Views of Ancient Bones," *Science
    News,* Vol. 122, August 1982, p. 116.)

Australopithecus were found at the same level. But then he also reminds us of the following discovery, originally noted by his father Louis Leakey, but thereafter mostly ignored.

> At one locality, remains of a stone structure—perhaps the base of a circular hut—were uncovered; there is an excellent date of 1.8 million years for this.[1]

Now a circular stone hut could hardly have been constructed by anyone but a true human being, but the stratigraphic level of this structure was *below* the levels of fossils of both *Australopithecus* and *Homo erectus!* And then, how about the remarkable intimations from the new science of cladistics, especially when correlated with the assumed chronology of continental drift? After showing that "cladograms" (diagrams of assemblages of physiologic and morphologic similarities, used to deduce relationships) of birds, butterflies, reptiles, and plants correspond globally with the chronology of splitting and drifting continents, two specialists in this field proceeded to show that cladograms of human beings, when classified according to race, language, and biochemistry, show exactly the same type of correlation. But the primeval super-continent is supposed to have split and initiated the development of these relationships about 80 million years ago. Consequently,

---

1. Richard Leakey, "Hominids in Africa," *American Scientist,* March/April 1976, p. 177.

these authors pose the following question:

> Would we not have to consider the
> possibility that humans also are that old,
> and have been affected by the same
> events.[1]

This would mean that man is not only as old
as the australopithecines, but also as old as
the dinosaurs. To the creationist, of course,
this is eminently reasonable. Many
"anomalous" human fossils and artifacts have
been reported throughout the geologic column,
but these have been commonly ignored or
explained away by evolutionists. The best case
in point is the well-documented case of overlapping dinosaur and human footprint trails in
Texas[2] which, despite an abundance of solid
evidence, is still being trivialized by
evolutionists.

But that is another story. The point we
emphasize here is that the fossil record of man
and the apes, like that of all other creatures, is
one that testifies of stability of the basic
created kinds, variation within the kinds,
occasional extinction of kinds, with clearcut

---

1. Gareth Nelson and Norman Platnick, *Systematics and
    Biogeography* (New York: Columbia University Press,
    1981). Of course, if evolutionists cannot be persuaded
    by fossil footprints, they will probably not be influenced
    by cladograms.
2. John D. Morris, *Tracking Those Incredible Dinosaurs and
    the People Who Knew Them* (San Diego: Creation-Life
    Publishers, 1980), 250 pp.

and permanent gaps between the kinds. There has never been any *real* evolution.

THE HOPELESS MONSTERS

All of which brings us to the amazing new concept known as *punctuated equilibrium*. This idea was introduced in this country only in 1971 by Niles Eldredge and Stephen Jay Gould (though it had previously been popular in Russia), but is now rapidly gaining dominance over the field as far as evolutionary theory is concerned. It is the product *not* of any evidence—but rather of *lack* of evidence—for evolution. Since the universal absence of transitional forms in both the fossil record and the living world had shown that slow-and-gradual evolution was invalid, and since the traditional mutation/selection mechanism of neo-Darwinism had proved impotent to generate anything of consequence, many evolutionists have finally decided they must resort to *revolutionary evolutionism.* The only other alternative seemed to be creationism, and that, to the leaders of evolutionary thought, was unthinkable heresy. So, instead of turning to God, they attributed the generation of each new kind of organism to some mysterious genetic upheaval, some remarkable embryonic saltation, some lucky leap of jumping genes, transforming a decadent population of organisms in equilibrium through a dynamic punctuation in that equilibrium to a new and higher degree of biologic existence. Thus, paradoxically, the main characteristic of evolution has become stability. Evolution,

which implies "change," is mostly
characterized by *stasis*, which means "no
change." The mechanism of evolution is long
periods of equilibrium punctuated by brief but
dynamic episodes of chance upheavals, which
somehow heave things up instead of down. One
of the leading advocates of this strange idea (in
addition to Gould and Eldredge) has been
paleontologist Steven Stanley, of Johns
Hopkins University. Note this remarkable
statement from one of his recent books:

> The (fossil) record now reveals that
> species typically survive for a hundred
> thousand generations, or even a million or
> more, without evolving very much. We
> seem forced to conclude that most
> evolution takes place rapidly, when
> species come into being by the
> evolutionary divergence of small
> populations from parent species. After
> their origins, most species undergo little
> evolution before becoming extinct.[1]

This is an amazing excurs into the logic of
wonderland. Since species survive indefinitely
without significant change (except extinction),
therefore they evolve very rapidly! Creationists,
of course, consistent with the testimony both
of divine revelation and of all the facts of
observable science, have long stressed this
stability of kinds ("after its kind" occurs ten

---

1.   Stephen M. Stanley, *The New Evolutionary Timetable:
      Fossils, Genes, and the Origin of Species* (New York:
      Basic Books, Inc., 1981), Preface.

times in Genesis Chapter One). This is exactly what the "creation model" would predict— primeval completed creation and subsequent conservation of the biologic kinds. But evolutionists reject creationism, not because of the testimony of science, but because the Bible teaches it.

Because of this innate rebellion against God, the modern school of evolutionists thus has turned to a theory for which there is no mechanism and no evidence. Stanley and others like to emphasize what they call "quantum speciation," since there is no evidence for gradual evolution, even at the species level. If the origin of new species is inexplicable, however, the origin of higher categories (genera, families, orders, etc.) is far more so. We have already quoted Gould on the impossibility of accounting for the origin of the Baupläne by any form of gradualism (see pp. 62,63).

The evolutionist has long used mutations as his explanation of the origin of new features in organisms. The main problem has always been that real mutations always seem to be harmful. In fact, individual creatures which experienced significant mutations were once commonly called "monsters," since they were seriously handicapped—if they survived at all.

In the 1930s and 1940s, two leading scientists had proposed that occasional "hopeful monsters," as they called them, must have been the means by which evolution had advanced. Europe's top paleontologist, O. H. Schindewolf, and Richard Goldschmidt, one of America's outstanding geneticists, vigorously promoted this idea, pointing out that everything known about genetics and paleontology showed that slow-and-gradual evolution had not occurred and could not occur. However, neo-Darwinism prevailed at the time and the hopeful monster idea was mostly ridiculed, despite the high reputations of its advocates.

Now, however, evolutionists finally have acknowledged that slow-and-gradual evolution really doesn't work after all, and the hopeful monster is being taken seriously. Gould has actually written a key article called "The Return of Hopeful Monsters" (*Natural History*, July 1977), predicting that Goldschmidt soon would be vindicated. Stanley also has referred to this concept in several places. For example, he illustrates it with a familiar example:

> Schindewolf believed that a single *Grossmutation* could instantaneously yield a form representing a new family or order of animals. This view engendered such visions as the first bird hatching from a reptile egg.[1]

Regardless of the terminology that may be preferred, however—hopeful monster, *Gross-*

---

1.   Stephen M. Stanley, *Macroevolution: Pattern and Process* (San Francisco: W. H. Freeman and Co., 1979), p. 35; see also p. 159.

mutation, quantum speciation, punctuated equilibrium, or whatever—the facts are still the same. There are no transitional forms and there is no known mechanism that will accomplish gradualistic evolution. Consequently, if one will not accept creation, he simply has to believe in some as-yet-undiscovered mechanism which will cause new and more complex organisms to evolve all at once—or at least so rapidly as to leave no record of the few intermediate steps. No such example has been recorded in all human history (stories of frogs turning into princes are not found in history books!), but there seems no other alternative.

Is the punctuated-equilibrium/hopeful-monster concept really a plausible option? It would seem that, if slow-and-gradual evolution must be rejected for lack of evidence, then sudden evolution must be even more vigorously rejected, for its evidence is not only missing, but inconceivable. One of the most revered of the neo-Darwinians, population geneticist Sewall Wright, has protested as follows:

> The reorganization required for the origin of the highest categories may seem so great that only "hopeful monsters" will do. Here, however, we must consider the size and complexity of the organisms. Such changes would probably have been impossible except in an organism of very small size and simple anatomy. I have recorded more than 100,000 newborn guinea pigs and have seen many hundreds of monsters of diverse sorts, but none were remotely "hopeful," all having died

shortly after birth if not earlier.[1]

Can anyone seriously believe that the first bird really hatched out of a reptile egg? Actually, there would have to be at least two such hopeful monsters—one male and one female—occurring simultaneously in the same population, in order to assure survival of the new type. It would seem that one could as easily believe in a fairy godmother with a magic wand!

Evolutionists sometimes insist that the punctuated equilibrium concept applies only to sudden change at the species level, and so should not be confused with the hopeful monster idea, which presumably would apply, if at all, only to higher categories. However, Gould and other punctuationists have frequently referred to the sharp gaps between phyla, classes, orders, and families as evidence in favor of punctuated equilibrium (note the quotations on pages 62, 65, 66, as typical of many examples that could be cited). In any event, the fact remains that there are no true transitional forms (with transitional structures) between these higher categories.

There is still considerable resistance[2] to this

---

1. Sewall Wright, "Character Change, Speciation, and the Higher Taxa," *Evolution*, Vol. 36, No. 3, 1982, p. 440.
2. G. L. Stebbins and F. J. Ayala say: "The specific solution postulated by Goldschmidt, that is, the occurrence of systemic mutations, yielding hopeful monsters, can be excluded in view of current genetic knowledge." (*Science*, Vol. 213, 1981, p. 969.)

punctuated equilibrium concept among evolutionists (as well there should be!), but it has probably become the dominant view. We have already noted the admitted difficulty of even imagining viable intermediate forms in the gradualistic evolutionary alternative, as well as the complete lack of fossil evidence for it. If evolutionists were really willing to take the scientific evidence seriously, they would have to conclude that *neither* gradual upward evolution nor saltational upward evolution has ever occurred and that neither ever could occur.

But this conclusion would make them creationists! They would then be acknowledging that God's record of creation was right after all, and this they refuse to do. When confronted with overwhelming evidence that Jesus Christ is Creator, by an act of self-will they reject Him! "We will not have this man to reign over us" (Luke 19:14), they say. And that is that.

Chapter 4

Evolution and Revolution

HOW TO CHANGE SOCIETY

It may not be just an accident of history that the punctuated equilibrium concept of the 1970s followed the radical student movements of the 1960s, for the one seems a natural out- growth of the other. The leaders of this new movement are almost all young scientists (Gould was born in 1941, Eldredge in 1943, Stanley in 1941, etc.) who were in graduate school in the sixties, whereas the leaders of neo-Darwinism were all scientists of the older generation (Simpson was born in 1902, Huxley in 1887, Dobzhansky in 1900, Mayr in 1904, Stebbins in 1906, etc.).

This older school was committed to gradualism, the gradual accumulation of very small beneficial mutations by natural selection, the survival of the fittest in the struggle for existence over long ages. Furthermore, the original founders and popularizers of Darwinian evolution had been products of nineteenth century capitalism, and they had more or less inevitably applied evolutionary theory to their society, as well, using it to justify laissez-faire capitalism, imperialism, and Caucasian racism. Herbert Spencer, the tremendously influential English evolutionary philosopher and sociologist, vigorously promoted the now-repudiated ideas of social Darwinism, which were applied with a vengeance by American industrialists such as John D. Rockefeller and Andrew Carnegie, and British imperialists such as Cecil Rhodes. In Germany, the racist evolutionism of men like Haeckel, Nietzsche, and Bismarck eventually produced the enormities of Nazism under the intense evolutionism of Adolph Hitler.

University students of the immediate post-war years (1945-1960) were imbued to a considerable degree with the patriotic fervor of the World War II period, as well as the Korean War and the anti-Communist Cold War. Economic prosperity seemed within reach of all and neo-Darwinism was in its heyday, reaching its zenith at the great Darwinian Centennial celebrations in 1959.

But the students of the 1960-70 decade were from a different background. They were the "war babies," born during or shortly after World War II, and were not nearly as enamored of patriotism or western-style evolutionary capitalism as their parents had been. They were not as willing to fight Communists in Asia as their parents had been to fight Nazis in Europe or their older brothers to fight Communists in Korea, because they had little understanding of the supposed cause for which they were fighting. In fact, as far as they could discern, the teachings of Marxism (on which Asian communism presumably was based) sounded rather noble and idealistic. At least, this was the way many of their teachers in high school and college had made it sound.

Despite what may have been the excesses of so-called McCarthyism during the Cold War period, there had indeed been much hidden—but effective—Marxist activity in America during those years, and this had at least succeeded in developing much influence in the schools, especially in such key departments as education and journalism, whose graduates would have the greatest influence in molding minds and shaping opinions in the future, and in such key universities as Harvard, M.I.T., Columbia, the University of California system, and others. Although there may not have been many actual Communist Party members involved, their dedicated activity began to wield great influence, and theoretical Marxism, at least, began to assume a high degree of acceptability in these institutions of higher learning. Much sympathy began to be generated for the anti-imperialist and anti-racist struggles of "the people" in the Third World countries. In our own country, the struggle for racial equality, for economic equality, and other noble causes were commingled with increasing resistance to fighting Vietnamese Communism, until all this exploded in the campus riots and the street violence of the late 1960s.

Now it should be noted that these student movements were not ostensibly fighting against creationism or religious fundamentalism. These were dead issues to them. They had all been indoctrinated in the "fact" of evolution and even the fundamentalist remnant, after the Scopes trial, had largely either accommodated evolution (via the day/age theory) or agreed to ignore it (via the gap theory). The question at issue was the *mechanism* of evolution. Increasingly it was becoming apparent that the traditional Darwinian and neo-Darwinian interpretations of evolution constituted the very rationale for the establishment evils they were fighting. Gradualistic, struggle-for-existence, survival-of-the-fittest evolution had been used to rationalize imperialism and racism, as well as economic and environmental exploitation of people and resources. They could not question evolution, which was a "proven fact of science," but perhaps a different mechanism of evolution should be devised, along with a different political and social system which would be more consistent with this alternative mechanism.

THE RACISM OF DARWINISM

The fact that Darwinian evolution is the basis of modern racism has been documented elsewhere by creationists,[1] but it is significant that this fact is finally being acknowledged—and even emphasized—by these modern evolutionists, after being indignantly denied for so long by older evolutionists whenever creationists would mention it. Stephen Jay Gould says, for example, speaking of the now-discredited recapitulation theory, which was long used as an argument for evolution:

---

1. *E.g., The Troubled Waters of Evolution,* by Henry M. Morris (San Diego: Creation-Life Publishers, 1975), pp. 40-46, 161-166.

> In Down's day, the theory of recapitula-
> tion embodied a biologist's best guide for
> the organization of life into sequences of
> higher and lower forms. (Both the theory
> and "ladder approach" to classification
> that it encouraged are, or should be,
> defunct today.) This theory, often ex-
> pressed by the mouthful "ontogeny re-
> capitulates phylogeny," held that higher
> animals, in their embryonic development,
> pass through a series of stages represent-
> ing, in proper sequence, the adult forms of
> ancestral, lower creatures . . . . Re-
> capitulation provided a convenient
> focus for the pervasive racism of white
> scientists . . . .[1]

Gould has rightly pointed out that this absurd
theory is now defunct, as it should be, but the
strange thing is that many people still believe it
(we encounter it frequently in creation/evolution
debates). Any lingering doubt should have been
eliminated by modern fetoscopy, which can actually monitor the fetal growth in the womb.

---

1. Stephen Jay Gould, "Dr. Down's Syndrome," *Natural
    History*, Vol. 89, April 1980, p. 144. The title of this arti-
    cle has reference to the physiologic infirmity widely
    known as mongolism, first described and named by Dr.
    Down. The idea was that the "races" had evolved as
    Negroid, Mongoloid, and Caucasoid, in that order, and
    that a mongoloid child, in his embryologic and infantile
    growth, had simply been arrested at that stage in his
    evolutionary recapitulation.

> Fetoscopy makes it possible to observe
> directly the unborn child through a tiny
> telescope inserted through the uterine
> wall . . . . The development of the child—
> from the union of the partners' cells to
> birth—has been studied exhaustively. As a
> result, long-held beliefs have been put to
> rest. We now know, for instance, that man,
> in his prenatal stages, does not go
> through the complete evolution of life—
> from a primitive single cell to a fishlike
> creature to man. Today it is known that
> every step in the fetal developmental proc-
> ess is specifically human.[1]

Nevertheless, as Gould said, this theory did provide in the 19th century, "the best guide for the
organization of life into sequences of higher and
lower forms." It was *not* the stratigraphic sequences, but the recapitulation theory popular-
ized by Ernst Haeckel, that gave 19th century
paleontologists their system for organizing their
fossils into an evolutionary series!

That the profound racism engendered by such
notions persisted well into the 20th century is
evidenced in a remarkable article by Henry Fairfield Osborn, probably the leading evolutionary
anthropologist of the first half of the 20th century, as well as president of the American
Museum of Natural History.

> If an unbiased zoologist were to descend
> upon the earth from Mars and study the

---

1. Sabine Schwabenthan, "Life Before Birth," *Parents*,
   October 1979, p. 50.

races of man with the same impartiality as
the races of fishes, birds, and mammals,
he would undoubtedly divide the existing
races of man into several genera and into
a very large number of species and sub-
species.[1]

Note the evolutionary use of the term "race"
here, a word (or concept) which never appears in
the Bible, despite the gratuitous contention of
some evolutionists that racism is derived from a
fundamentalist interpretation of Scripture. Osborn then says:

This is the recognition that the genus
Homo is subdivided into three absolutely
distinct stocks, which in zoology would be
given the rank of species, if not of genera,
stocks popularly known as the Caucasian,
the Mongolian, and the Negroid.
The spiritual, intellectual, moral, and
physical characters which separate these
three great human stocks are far more
profound and ancient than those which
divide the Nordic, Alpine, and Mediterran-
ean races. In my opinion, these three
primary stocks diverged from each other
before the beginning of the Pleistocene or
Ice Age . . . . The standard of intelligence
of the average adult Negro is similar to
that of the eleven-year-old youth of the
species *Homo sapiens*.[2]

---

1.   Henry Fairfield Osborn, "The Evolution of Human
     Races," *Natural History*, Jan/Feb 1926, reprinted in
     Natural History, Vol. 89, April 1980, p. 129.
2.   *Ibid.*

So the punctuational evolutionists have made their point that Darwinism has led to racism. They have also stressed its culpability with respect to social Darwinism (exploitative capitalism, imperialism, etc.), but this has long been acknowledged, with regrets, even by neo-Darwinites. They are not so eager to acknowledge its responsibility for Fascism and Nazism (after all, Mussolini and Hitler did call their movements "socialistic," and the student movements of the 1960s bore many striking resemblances to the Hitler youth of the 1930s and early 1940s), but the connection is certainly there.

> Hitler believed in struggle as a Darwinian principle of human life that forced every people to try to dominate all others; without struggle they would rot and perish . . . . Even in his defeat in April 1945 Hitler expressed his faith in the survival of the stronger and declared the Slavic peoples to have proven themselves the stronger.[1]

Which brings us, then, to a consideration of evolution as it is understood by these Slavic peoples, the Russians in particular, for it is *this* model which seems to have captivated the hearts of the modern school of western evolutionists.

MARXIST PUNCTUATIONISM

Karl Marx, of course, was a committed evolutionist before Darwin published *The Origin of*

---

1. P. Hoffman, *Hitler's Personal Security* (London: Pergamon Press, 1979), p. 264.

Species, but he was such because of his atheistic faith (born a Jew, he had once made a Christian confession, presumably for economic reasons, but later renounced all organized religions in favor of atheism) and his commitment to change in human history and economic systems, but he had no real scientific basis for this faith until Darwin came along. As is well known, Marx even wanted to dedicate *Das Kapital* to Darwin, but Darwin refused.

In spite of Marxist commitment to evolution, however, both Marx himself and later Communists were always uncomfortable with the uniformitarian aspects of Darwinism. They honored Darwin because of the scientific respectability which he had given to naturalism and atheism[1] but they needed a more immediate mechanism for evolutionary change than the ideas of long-drawn-out progress which commended themselves to western Europe and America, the nations of which were experiencing unprecedented growth and development.

Lamarckianism was very appealing to the communist mind, for it promised evolution through changing the environment. This could be a considerably more rapid process than the cut-and-try

---

1. "And of all those eminent researchers of the nineteenth century who have left us such a rich heritage of knowledge, we are especially grateful to Charles Darwin for opening our way to evolutionary dialectical understanding of nature." (Cliff Conner, "Evolution vs. Creationism: In Defense of Scientific Thinking," *International Socialist Review: Monthly Magazine Supplement to the Militant,* November 1980.)

mechanism of natural selection. A revolutionary change imposed on the environment could produce all sorts of physical, as well as social, benefits to society, through the Lamarckian inheritability of characters produced environmentally. For a considerable time, therefore, this concept was even imposed on Russian scientists as official state dogma.

The problem, of course, was that Lamarckian evolution—unlike Darwinian tautologies—was testable! It was, indeed, subjected to many scientific experiments, and, in spite of optimistic claims made for awhile by such Russian scientists as Kammerer, Pavlov, Lysenko, and others, it was eventually so completely falsified by tests that it finally had to be officially repudiated.

This did not mean, however, that they would simply capitulate to neo-Darwinism, for this was the quasi-official dogma of western capitalism and, besides, it acted too slowly. Communism must proceed by *revolution,* not slow-and-gradual *evolution!* To be scientific, some form of naturalistic mechanism had to be found which would elevate systems rapidly—not gradually—to a higher state. And so, apparently, was born the complex of concepts which we have called *revolutionary evolutionism.*

Steven Stanley has described the introduction of this system of evolutionary thought to North America as follows:

> The punctuational idea emerged as a more visible alternative to English-speaking paleontologists with the publications of Eldredge (1971) and Eldredge and Gould (1972). It is both interesting and

surprising that, unknown to Americans,
this view had previously gained support in
the paleontologic community of the Soviet
Union.[1]

Gould and Eldredge have themselves admitted
this Marxist pedigree for their theory. First,
however, they present a philosophical justification:

> Alternative conceptions of change have
> respectable pedigrees in philosophy.
> Hegel's dialectical laws, translated into a
> materialist context, have become the of-
> ficial "state philosophy" of many socialist
> nations. These laws of change are explicit-
> ly punctuational, as befits a theory of
> revolutionary transformation in human
> society.[2]

These authors thus stress that Hegel's dialectical
materialism, which was adopted by Marx as Communist dogma when put in the context of
materialistic evolution, is the state "philosophy"
(read "state religion") of those nations which
have become Communist (or "socialist").

Another geologist seemingly sympathetic to
Marxist "science" has commented similarly, as
follows:

> The theory of dialectic materialism

---

1. Steven M. Stanley, *Macroevolution: Pattern and Process*
    (San Francisco: W. H. Freeman and Co., 1979), p. 36.
2. Stephen J. Gould and Niles Eldredge, "Punctuated
    Equilibria: The Tempo and Mode of Evolution Recon-
    sidered," *Paleobiology*, Vol. 3, Spring 1977, p. 145.

> postulates matter as the ultimate reality,
> not to be questioned . . . . Evolution is
> more than a useful biologic concept: it is a
> natural law controlling the history of all
> phenomena.[1]

This writer proceeds to justify the use of circular reasoning in geologic dating and other interpretations, denying any real significance to so-called "absolute time" (so important in uniformitarian neo-Darwinism) on the basis of this materialistic premise, saying in effect that all that counts is pragmatism—that is, whether a system works out or not.

Gould and Eldredge apparently believe that Marxist philosophy will work and are quite pleased to be able to replace the capitalistic system of evolution with the Marxist brand.

> In the light of this official philosophy, it
> is not at all surprising that a punctua-
> tional view of speciation, much like our
> own, but devoid (so far as we can tell) of
> references to synthetic evolutionary theory
> and the allopatric model, has long been
> favored by many Russian paleontologists.
> It may also not be irrelevant to our per-
> sonal preferences that one of us learned
> his Marxism, literally, at his daddy's knee.[2]

It is interesting that Gould has occasionally

---

1. J. E. O'Rourke, "Pragmatism versus Materialism in Stratigraphy," *American Journal of Science,* Vol. 276, January 1976, p. 51.
2. Gould and Eldredge, *op cit,* p. 146.

been embarrassed by his Marxist reputation and has waffled about it somewhat. On other occasions, however (at least once when under oath), he has acknowledged that he is, indeed, a Marxist, and he is generally recognized as such by his colleagues (there are many other Marxists in the academic world today, especially at Harvard, M.I.T., and other "prestige" schools). A recent reference to Gould in the "theoretical and discussion journal of the British Communist Party" says, however, that he is *not* a Marxist.

> The most eloquent defender of scientific evolutionism, Stephen Jay Gould, is an avowedly non-Marxist radical—on the left of the scientific/political consensus, but working well and truly within it.[1]

It would, no doubt, in the current situation, serve the interests of doctrinaire Marxism best for such a prominent scientific evolutionist as Gould *not* to be identified officially as a Marxist, whatever his actual beliefs may be. This article also has the following to say about evolution in general:

> Aspects of evolutionism are perfectly consistent with Marxism. The explanation of the origins of humankind and of mind by purely natural forces was, and remains, as welcome to Marxists as to any other secularists. The sources of value and responsibility are not to be found in a

---

1. Robert M. Young, "The Darwin Debate," *Marxism Today,* (Vol. 26, April 1982, p. 22.

separate mental realm or in an immortal
soul, much less in the inspired words of
the Bible.[1]

In any case, the introduction of punctuationism
into American evolutionary thinking just a
decade ago has rolled almost like a tidal wave
over the whole field, especially the younger
generation of evolutionists, disenamored with the
capitalistic establishment and seeking social
justice and full egalitarianism not by slow
evolutionary change but by rapid and even
violent change if need be. Gould and his colleagues
have been lionized as leaders of this new wave of
revolutionary science. Gould (as well as Carl
Sagan, Carl Johanson, and other young
evolutionists) are personable, intelligent,
articulate, and extremely popular. Gould has
been featured on the cover of *Newsweek,* named
"Man-of-the-Year in Science" by *Discover,* has
appeared as guest star on the "Donahue Show"
and other national telecasts, and has received
great adulation. Sagan (with his super-expensive
Cosmos series on the Public Broadcasting
System), Johanson (with his famous "Lucy"
hominid—named after a spaced-out character in
a Beatle's song), and others have received
almost as much notoriety in the feverish
promotion of evolutionism by the news media in
recent years.

---

1. *Ibid,* p. 21.

NATURALISTIC CATASTROPHISM

The new wave of scientific revolutionism has not by any means been limited to paleontology and biology, of course. We have already referred to the chaotic state of modern genetics and the wistful search for a genetic mechanism that can produce hopeful monsters. Likewise the field of geology, long committed to uniformitarianism, is experiencing a great revival of catastrophism in geologic interpretation. The geologic literature is full, these days, of cometary and meteoritic impacts, cosmic ray bombardments resulting from magnetic field reversals or nearby super-novas, ancient nuclear explosions, gigantic earth movements, floods, volcanic explosions, land-slides, and a wide variety of other local and regional—sometimes even global—catastrophes.

This is not a return to Biblical catastrophism, of course. It is only because the older ideas of Biblical geology, centered around the Genesis flood, were considered buried that modern geologists have felt free to reintroduce catastrophism into their interpretations.

And, not surprisingly, these catastrophes are more and more being synchronized, ideologically at least, with periods of explosive evolution and extinction—elimination of the older, unfit animals, such as the dinosaurs, with the rapid evolution of new and better animals, for the new age. Somehow the chaotic environment of the catastrophe triggers the long-dormant evolution-ary urge within the genetic system and, lo and behold, new and higher kinds appear.

Even the classical field of astronomy has been invaded by catastrophism. Not only do we now have great emphasis on the very rare events known as novas and supernovas, as well as such esoteric concepts as black holes, anti-matter, time reversals, and the like, but the universe itself is commonly believed to have originated in the greatest punctuational catastrophe of all, the primordial Big Bang. If the cosmos started in such a great upheaval, it should be no great thing to anticipate periodic subsequent revolutions in which still further evolutionary advances can be generated.

However, there is one overwhelming problem with this whole scenario of increasing organization by big bangs, supernovas, punctuational catastrophes, quantum speciations, and hopeful monsters. In the real world of observational science, things never work this way! Explosions and catastrophes generate disorganization, not organization, and monsters are never hopeful.

The Second Law of Thermodynamics, which has been discussed repeatedly by creationists[1,2,3] is an impregnable barrier to such evolutionary advances (without miraculous intervention, at least). If this universal principle of increasing disorder is a problem to neo-Darwinism (and it assuredly *is* a problem, evolutionary wishful think-

1. Henry M. Morris, *Scientific Creationism* (San Diego: Creation-Life Publishers, 1974), pp. 37-58.
2. Creation Research Society, *Thermodynamics and the Development of Order* (Edited by Emmett L. Williams, Creation Research Society Books, 1981).
3. Henry M. Morris and Gary E. Parker, *What is Creation Science?* (San Diego: Creation-Life Publishers, 1982), pp. 154-188.

ing to the contrary notwithstanding), then it is a problem much compounded for punctuationism. If small random mutations produce decreased organization (as they must and do), then big random changes will produce much more severe disorganization. Little mutations are *harmful*—big mutations are *lethal.* Or at least that is how it is in the real world of experience and observation.

But modern revolutionary evolutionists have an answer (really only a pseudo-answer) even to this insuperable problem. This is the very remarkable concept (actually, it helped win for its fabricator, Ilya Prigogine, a Nobel prize in 1977) of *dissipative structures.* That is, somehow in the violent context of a high flow-through of energy, with a high degree of dissipation of that energy, a certain amount of "order" is produced. As heat flows through a pot of water, with its energy being largely dissipated in the process, certain ordered structures known as vortices and bubbles are produced for awhile. As the solar radiation flows by the earth, some of it generates atmospheric motions which produce ordered structures known as hurricanes and tornados.

Even though these bubbles and hurricanes generate still more disorder and energy dissipation, they do have some degree of "order" for a little while, so they are called "dissipative structures." Prigogine also used the term "order through perturbations." The hope is, of course, that the order in dissipative structures could somehow be retained and used as a foundation for a still higher degree of order when the next wave of highly dissipating energy flows past. Evolutionists hope that this idea can eventually be used to explain the origin of living structures from nonliving chemicals in some high-energy field and then, perhaps, more complex living structures in an active field of jumping genes or other genetic turmoil.

The political application is obvious, even though scientific proof is certainly not yet at hand for any such unscientific concept.

> Prigogine's work has long been of interest to systems theorists seeking to apply the logic of their fields to global problems. One such scientist is Ervin Laszlo of the United Nations. "What I see Prigogine doing," says Laszlo, "is giving legitimization to the process of evolution—self-organization under conditions of change. . . . Its analogy to social systems and evolution should be very fruitful."[1]

Now this all sounds far-fetched, and it is, but we have heard evolutionists maintain over and over again in recent years that Prigogine had solved the entropy problem as far as evolution was concerned. He had "given legitimization to the process of evolution," as Lepkowski says (an interesting admission, incidentally, that before Prigogine, evolution was illegitimate!). Somehow they think that these highly ephemeral and relatively trivial entities provide the directing code and energy conversion mechanism necessary to provide a continual age-long increase in organized complexity of the earth's open-system biosphere. The very idea is absurd.

---

1. Wil Lepkowski, "The Social Thermodynamics of Ilya Prigogine," *Chemical and Engineering News*, Vol. 57, April 16, 1979, p. 30.

No, Prigogine's dissipative structures have not legitimized evolution, but they may well serve as a pseudo-legitimization in the minds of young intellectuals for social action in line with the imaginary inferred analogy with past evolution. That is, if evolutionary progress has actually been achieved in a milieu of high energy dissipation (violent revolution) and catastrophe (extinction of the reactionary dinosaurs that dominate the present age to make way for the swift runners and high fliers of the new age), then why not get on with it!

There will be resistance, of course, so the more "scientific" such notions can be made to appear, the sooner they will be accepted. The conflict was not expected to be between creationists and evolutionists, as noted before, but between the *avant garde* evolutionists and the "old-guard" evolutionists. Both groups are equally opposed to creationism and equally committeed to a completely naturalistic view of the world, but the "new wave" favors socialism, while the "old guard" clings to capitalism.

One of the leaders in the latter group is Beverly Halstead, Professor of Zoology and Geology at Reading University in England. Halstead has been waging a battle against the newer views that have been introduced recently at the British Museum, charging that they are Communist-oriented and aimed at revolution.

> The theoretical basis of communism is dialectical materialism, which was expounded with great clarity by Frederick Engels in *Anti-Duhrung and The Dialectics of Nature.* He recognized the great value of the contributions made by biology in

establishing that there was constant
movement and change in nature and the
significance of Darwin's demonstration
that this applied also to the organic world.
. . . The crux of the entire theoretical
framework, however, is in the nature of
qualitative changes. This is also spelt out
by Engels in *The Dialectics of Nature*, "a
development in which the qualitative
changes occur not gradually but rapidly
and abruptly, taking the form of a leap
from one state to another." . . . Here, then,
is the recipe for revolution.[1]

Nature magazine thereupon proceeded to pub-
lish a series of articles and letters on the subject.
A particular peeve of Halstead's was the Natural
History Museum's new emphasis on cladistics,
which attempted to group animals by physical
characteristics, rather than by evolutionary
ancestry, implying that each "clade" had to appear abruptly, without leaving evidence of its
ancestry in the fossils.

According to the stated assumptions of
cladistics, none of the fossil species can
be ancestral, by definition. This presents
the public for the first time with the notion that there are no fossils directly ante-
cedent to man. What the creationists have
insisted on for years is now being openly
advertised by the Natural History
Museum.[2]

---

1.   Beverly Halstead, "Popper-Good Philosophy, Bad
     Science?" *New Scientist*, Vol. 87, July 17, 1980, p. 215.
2.   L. Beverly Halstead, "Museum of Errors," *Nature*, Vol.
     288, November 20, 1980, p. 208.

Dr. Halstead did not mention that he had recently had an unfortunate encounter with Dr. Duane Gish, of ICR, in a debate at Reading University. The museum had been unable to provide him with any transitional fossils to use in the debate, and he suddenly realized that the cladistic punctuationism which was taking over the Museum was using the same arguments and evidences as the creationists, even though their motives were vastly different. He also learned that neo-Darwinian gradualism was under serious attack. At Harvard, there had even been a near-riot between followers of the two camps, as the Marxists took violent exception to the racist implications of "sociobiology," which is an application of neo-Darwinism in human society. Halstead said:

> The next question is why should the notion of gradualism arouse passions of such intensity. The answer to this is to be found in the political arena . . . . If it could be established that the pattern of evolution is a saltatory one after all, then at long last the Marxists would indeed be able to claim that the theoretical basis of their approach was supported by scientific evidence. Just as there are "scientific" creationists seeking to falsify the concept of gradual change through time in favor of creationism, so too there are the Marxists who for different motives are equally concerned to discredit gradualism.[1]

So the creationists have suddenly and in-

---

1.   *Ibid.*

advertently appeared in the middle of this conflict between the capitalist evolutionists and the
Marxist evolutionists. The ironical nature of this
development has been noted by Tom Bethell,
popular journalist with *Harper's* magazine.

> The left-wing critique of Darwinian
> theory has by no means prevailed, but if it
> should do so, let us also enjoy the fantas-
> tic irony that the fundamentalists, who
> have been trying for more than a hundred
> years to knock Darwin off his pedestal,
> without success, will be indebted not to
> the right-wingers, with whom they have
> always been aligned, but to biologists
> whose god is Marx.[1]

Creationists, of course, recognize that both
groups of evolutionists are right. That is, there
is no evidence for slow-and-gradual evolution,
as the punctuationists have shown, and there
is no evidence for catastrophic evolution, as
the gradualists have shown. There is no
evidence for any kind of evolution. Period.

---

1.   Tom Bethell, "Burning Darwin to Save Marx," *Harper's*,
     December 1978, p. 92.

Chapter 5

Anti-Creationist Hysteria

THE RETURN OF CREATIONISM

Into all this evolutionary turmoil suddenly entered an unexpected and highly unwelcome intruder. Evolutionists might argue interminably among themselves about the mechanisms and applications of evolution, but evolution itself must never be questioned! The furor and hysteria unleashed in the evolutionary establishment when creationism suddenly became a serious threat have been marvelous to behold.

There has always been a creationist remnant, of course, but for most of this century it had been discounted and largely ignored. Especially since the infamous Scopes trial (1925), it had been widely claimed and believed that all scientists were evolutionists. Evolutionary teaching has almost completely controlled the colleges and universities and, to only a slightly less degree, the high schools and elementary schools of our country, for over half a century. In fact, evolutionary philosophy has really been dominant, except for the brief creationist revival associated with the fundamentalist movement of the early years of the 20th century, ever since Darwin.

The creationists were ignored during those years because evolutionists generally considered them all to be either Bible-thumping fundamentalist preachers or uneducated laymen, constituting only an insignificant minority. As a matter of fact, even the Christian leadership—not just liberals, but evangelicals and fundamentalists, as well—were so intimidated by these humanistic pressures that they were increasingly trying to find ways to accommodate evolutionism in their respective educational and exegetical systems.

In the world of secular education, evolutionary philosophy was not only dominant in biology teaching, but throughout the whole curriculum. The natural sciences were based on evolutionary naturalism, the social sciences emphasized evolutionary socialism, the humanities stressed evolutionary humanism, and even the business and technology courses imbued a spirit of evolutionary materialism. The vital truth of a sovereign Creator God controlling the world and human life was either denied or ignored in all aspects of education.

As atheistic evolution was taught in secular colleges, theistic evolution and the allegorical theory of creation were taught in liberal religious schools such as those of the "mainline" denominations, the evangelical colleges developed the compromise of progressive creation, with its day-age theory of the Genesis record, and the fundamentalist schools settled on what might be called "irrelevant creation," concentrating on evangelism and spirituality while ignoring all "worldly" and controversial issues such as science and economics and pigeon-holing the evolutionary ages in a supposed "gap" in the early verses of Genesis.

Humanistic evolutionary thought had also come largely to control the news media and government. And all this was happening not only in America, but also in practically all other nations. Evolution had, to all intents and purposes, become the god of this world.

At this point, creationism suddenly began to surface again. Only this time it could not merely be dismissed as a prejudice of reactionary preachers or laymen. It was a movement led by scientists, and by scientists who were at least as conversant with all the relevant scientific data as were the evolutionists. They were attacking evolution and promoting creation—not only on Biblical grounds (for the instruction of their compromising Christian brethren)—but even more powerfully on scientific grounds, using the data of paleontology and geology, biochemistry and thermodynamics, statistics and mathematics, to demonstrate that creation was a much sounder scientific model of origins than evolution.

It is not our purpose in this chapter to trace the history of the modern creationist movement, although the writer hopes to prepare such a volume in the near future. Certain key events were the publication of *The*

Genesis Flood[1] in 1961, the formation of the Creation Research Society in 1963,[2] and the establishment of the Institute for Creation Research in 1970.[3] There are now many other creationist organizations—local, regional, national, and international—and scores of books on scientific creationism have been published.

WHY DO CREATIONISTS WIN DEBATES?

Probably the phenomenon which has had the greatest single impact on the evolutionary establishment, however, has been the series of creation/evolution debates held in the past ten years. These were debates on leading college and university campuses between scientists on the ICR staff and on the respective campuses. The debates were all supposed to deal strictly

---

1. John C. Whitcomb and Henry M. Morris, *The Genesis Flood* (Philadelphia: Presbyterian and Reformed Publ. Co., 1961), 518 pp. Many observers have suggested that this book was the catalyst which stimulated the creationist revival.
2. The Creation Research Society is strictly a membership organization of creationist scientists, all with postgraduate degrees in the natural sciences and firmly commited to Biblical Christianity and scientific creationism. It began in 1963 with 10 members and now has over 700 fully qualified scientists in its membership, publishing an excellent quarterly journal of research studies in creationism. Membership applications can be obtained from the Membership Secretary, in care of Concordia College in Ann Arbor, Michigan.
3. The Institute for Creation Research was originally the research division of Christian Heritage College, which

with the scientific evidence, ignoring the religious, sociological, and political dimensions of the creation/evolution conflict (more often than not, however, the evolutionist debater would *not* stick to scientific questions, but would insist on attacking the Bible and the religious beliefs of creationists, while simultaneously also pushing evolution on religious grounds). Over 130 such debates have been held to date, with audiences often numbering in the thousands and consisting mostly of students on the respective campuses, almost all of them thoroughly exposed previously to evolutionism, but few having ever heard before the scientific arguments for creation. The effects on these audiences have been profound, and the impact of the debates has been felt far beyond the campus scene.

The impact of the debates can best be

---

was also established in 1970, as a uniquely creationist liberal arts college. Reorganized in 1972 under its present name, the ICR has no "memberships" as such, but an actual staff of scientists and support personnel, engaged in a wide-ranging program of research, writing, and teaching in the broad fields of Biblical and scientific creationism. It has produced over 60 books, as well as many audiovisuals, has a worldwide weekly radio program, and has given seminars, lectures, and institutes in many hundreds of schools, churches, and other locations around the world. ICR became an independent transdenominational educational institution in 1980 and now also offers state-approved graduate degree programs in the sciences and science education—so far as known, the only fully creationist and Biblically evangelical institution in the world which is doing so.

gauged by the concern shown by the evolutionists. A letter to the editor of *Bioscience* is indicative.

> Why do creationists seem to be the consistent winners in public debates with evolutionists? . . . We biologists are our own worst enemies in the creationist-evolutionist controversies. We must no longer duck this and other issues related to biology and human affairs, and when we do face them we must think clearly and express ourselves accordingly. We may still not be consistent winners in the creationist-evolutionist debates, but let the losses that occur be attributable to other than lapses in professional standards.[1]

Much more vitriolic is a recent article by Joel Cracraft, of the Department of Anatomy at the University of Illinois Medical Center in Chicago. Dr. Cracraft was a science advisor to the American Civil Liberties Union at the Arkansas Creation Trial of 1981, as well as organizer of an anti-creationist symposium at the 1981 convention of the American Association for Advancement of Science. Note, in the following, his back-handed admission of the rising influence of creationism and the poor case that can be made for scientific evolutionism.

---

1. Dennis Dubay, "Evolution/Creation Debate," *Bioscience,* Vol. 30, January 1980, pp. 4-5.

With their pleas for a fair hearing, the
scientific creationists, in particular those
at the ICR, excel in duping the public and
manipulating both the press and their
adversaries. How else can one explain
their countless speaking invitations from
secular universities and the willingness of
many scientists to debate them, as if
"scientific" creationism were an
intellectual equal of evolutionary biology
or, more accurately, science as a whole?[1]

One could possibly explain this, Dr. Cracraft,
by the fact that more and more people,
especially college students, are becoming
aware that they have, indeed, been duped for
many years, though not by creationists. As Dr.
Colin Patterson, one of the world's top
evolutionists, said recently:

Then I woke up and realized that all my
life I had been duped into taking
evolutionism as revealed truth in some
way.[2]

It is most significant that Cracraft expresses
strong objection to debating evolution and
creation on their scientific merits:

Debates are to the advantage of the
creationists especially when they create
the ground rules: "We will only debate the

---

1. Joel Cracraft, "Reflections on the Arkansas Creation
    Trial," *Paleobiology*, Vol. 8, No. 2, 1982, p. 83.
2. Colin Patterson, "Evolution and Creationism," Speech
    at American Museum of Natural History, New York, Nov.
    5, 1981. Transcript, p. 2.

> scientific evidences . . . religion and
> philosophy will not be subjects for
> discussion." . . . Structured in this way,
> debates invariably place scientists on the
> defensive because the creationists seek to
> promote a dualistic philosophy: evidence
> against evolution, they argue is evidence
> for creationism. On the other hand, it is no
> surprise that creationists are reluctant to
> debate an altogether different proposition.
> Is "scientific" creationism science or
> religion?[1]

In other words, evolutionists would much
rather attack the religious aspects of
creationism than defend the scientific aspects
of evolutionism! Creationists find it very
difficult to understand why, if evolution is a
proven fact of science (as evolutionists
continually insist, and as they perpetually
brainwash students into believing), a debate
argued strictly from the scientific evidence
would be "to the advantage of the
creationists."

Concerning Cracraft's petty lament that
creationists present only negative evidence
against evolution, surely he is intelligent
enough to realize that evidence against
evolution, such as the creationists present in
their lectures and debates, is, indeed, positive
evidence for creation. *There are only these two
possibilities!* Either all things have developed
by natural processes in a self-contained

---

1.   Joel Cracraft, *op cit*, pp. 83-84.

universe, or at least some things have come by supernatural processes in an open universe. The one alternative is evolution, the other is creation, and there is no third option. In this type of debate, the creationist is not arguing against a particular form of evolution (e.g., neo-Darwinism) or for a particular account of creation (e.g., the Genesis record), but only the broad and basic question of whether observed scientific data fit evolution better than they fit creation. For example, the evolution model would be strongly confirmed if one could find true transitional structures between major kinds of organisms, either in the fossil record or in living organisms (e.g., "sceathers," reptilian scales in the process of evolution into avian feathers). The creation model indicates no such thing ever existed, since the Creator would have created fully functional structures for each creature in accordance with His intended purpose for that creature. Thus, the ubiquitous, unbridgeable gaps between major kinds of organisms constitute strong evidence (though not "proof") against evolution and strong evidence (though not "proof") for creation. The same "dualism" (as Cracraft dubs it) will be found to apply to the other scientific arguments used by creationists (the second law of thermodynamics, the deteriorative nature of mutations, the improbability of complex or symbiotic systems, the conservative character of natural selection, etc., etc.).

The debates have surely brought one important fact to the attention of the public. Creationism can at least be discussed scientifically, strictly on the evidence from scientific data. Evolutionists may complain that creationists have a "hidden agenda," as the title of an article in the journal *Christianity Today* put it, and that they really believe in the Biblical account of creation even though they don't talk about it in their debates. But this is all irrelevant to the basic question of whether the actual hard data of paleontology, genetics, thermodynamics, etc., fit in with the *scientific* model of creationism—*as well as or better than they fit in with evolution.* This is the only aspect of creationism which creationists believe should be taught in public institutions anyway, so this is the only aspect which should be debated in such institutions.

But evolutionists, intuitively realizing that any real scientific case for evolution is very weak (creationists would say it is nonexistent!), prefer to defend it by branding creation as "religious" and then having creationism excluded from public institutions on the basis of church/state separation. A favorite tactic is to cite some creationist book or article expounding Biblical creationism and then to recoil with horror at the prospect of having *that* concept taught to public school children. Cracraft, for example, uses this device. After quoting (out of context) one of this writer's discussions on the subject of the possible influence of Satan, angels, and demons on earth and cosmic history (which are topics definitely treated in the Bible and even discussed in some detail by our Lord Jesus Christ) in a book specifically stated to be a Bible commentary and certainly not ever intended for use in public schools, Cracraft then makes the following utterly irrelevant and misleading comment:

> Inasmuch as Dr. Morris and others like him want to exercise control over the public school curricula, those reading this should take it upon themselves to make Dr. Morris' judgments about science available to local school boards, religious leaders, and state legislators. After all, he is Director of the most important creationist organization in the United States and will be directly involved in writing educational materials for the public schools if creationists get their way.[1]

Well, apparently all is fair in love and war, and evolutionists have definitely mounted a new war against creation.

THE BATTLE FOR THE PUBLIC SCHOOL

Creationists in no way wish to *control* public school curricula, of course, but they *have* been urging school administrators to be fair and to use a "two-model" approach in the schools. Since there are only two real world views of origins, history and meaning (that is, God-centered or man-centered, creationism or evolutionism, theism or humanism), why should tax-supported institutions be used to brainwash impressionable students only in evolutionary humanism? The latter is certainly no more scientific nor less religious than

---

1.   Joel Cracraft, *op cit,* p. 88.

theistic creationism.

Furthermore, the founding principles of our nation were based upon the truth of a Creator and of a natural order and moral law created by God. The Declaration of Independence itself begins with this premise. Dr. Elias Boudinot, president of the Continental Congress in 1782, in an Independence Day address that year, gave expression to what was commonly believed by Americans at that time:

> The history of the world, as well sacred as profane, bears witness to the use and importance of setting apart a day as a memorial of great events, whether of a religious or a political nature. No sooner had the great Creator of the heavens and the earth finished His almighty work, and pronounced all very good, but He set apart (not an anniversary, or one day in a year, but one day in seven, for the commemoration of His inimitable power in producing all things out of nothing.[1]

How, then, could it now be wrong to teach in our schools—at least as an alternative—the very premise upon which our nation and its government were founded? Dr. Boudinot, and most other American leaders of that day, even believed and taught *Biblical* creationism (he was, in fact, also the first president of the

---

1.   Elias Boudinot, "Independence Day Address," presented to the New Jersey Society of the Cincinnati, July 4, 1873. Cited in *Foundation for Christian Self-Government Newsletter*, July 1982, p. 3.

American Bible Society). Some, as deists, did
not believe in Biblical inerrancy, but they did
believe in God and creation. Even such a deist
as Thomas Jefferson, often quoted as the great
advocate of church-state separation, believed
in special creation and rejected the
evolutionary theories of his day. Separation of
church and state, incidentally, has never been
a constitutional requirement, only a modern
judicial interpretation. The First Amendment
prohibited an "established religion" which, in
the context of the times, meant only the state
endorsement and support of a particular sect
or denomination. It was certainly *never*
intended to ban God from the schools and to
establish the religion of secular humanism in
our schools, as has been done.[1]

For that matter, hardly anyone would even
argue that such was the original intent. The
problem is that evolution has been applied to
our legal system as well. Modern lawyers and
judges, steeped in evolutionary philosophy in
their training, have come to regard the
Constitution as "evolving" along with society.
There are no "absolutes" in an evolving world,
so even the Constitution, as well as the Bible
and the concept of God, must evolve along with
the changing mores of the current times.
Accordingly, "liberal" justices have gradually

---

1. Wendell R. Bird, "Freedom from Establishment and
    Unneutrality in Public School Instruction and Religious
    School Regulation," *Harvard Journal of Law and Public
    Policy,* June 1979, pp. 127-129, 138-140.

banned not only sectarianism, but even
Christianity itself from the classrooms of what
the founding fathers had intended (and the
Supreme Court had affirmed, more than once)
to be a Christian nation, one that accepted the
Bible as the authoritative Word of God. In the
process, of course, classroom prayer and Bible
reading have now also been outlawed in the
schools.

To date, however, the Court has never gone
so far as to ban from the schools the teaching
that there is a Creator God who created all
things in the beginning. To do so would, in
effect, officially establish the United States as
an atheistic nation, no different from Russia or
other Communist countries. The right to
believe and teach the truth of God and creation
is the irreducible minimum in the schools of
any nation which really holds to the principle
of religious freedom. Anything less will quickly
deteriorate into a secularized, humanistic,
atheistic dictatorship. Such a right need not
deny the freedom of secularists or humanists
or atheists to teach their religion, of course,
but there is not the slightest justification for
making their religion the *only* view that can be
taught.

But even though the Court has never banned
theistic creationism from the schools, a
multitude of bureaucrats and educational
administrators have interpreted Court decisions
that way. Creationist students and teachers
have been intimidated for years, and a number
of creationist teachers have even lost their
jobs.

It had been the hope of the creationist

scientists that, by stressing the scientific validity of creationism and making it clear that they were not asking for the Biblical version of creation to be taught in the schools, many teachers and school administrators would begin to take advantage of their constitutional rights (as well as those of their students), using the scientific evidences and arguments which were now becoming more accessible, to begin a real "two-model" approach in the schools, thus being fair to both evolutionists and creationists among their constituents.

Much headway was being made along these lines, and more and more schools and teachers began to do just this. The various teacher education programs of ICR, along with the literature now available, were becoming widely effective.

But the opposition began to grow as well, and progress seemed too slow to suit many impatient creationist activists. Deciding their constitutional rights were being violated (as indeed they were), many decided to try the political route, initiating litigation and/or legislation to *compel* the schools and other public institutions to adopt a two-model approach. Lawsuits were filed in Washington, D.C., Houston, Sacramento, and other places and creationist bills were introduced in many legislative assemblies.

Now, despite widespread publicity to the contrary, the Institute for Creation Research has always tried to discourage a legalistic and political approach to this issue (as has the Creation Research Society). History shows that neither scientific nor religious principles can be effectively legislated, and since there had been no legal restriction against teaching creation anyway, most creationist scientists have felt rather strongly that, in the long run, education and persuasion would accomplish more than legislation and coercion. Furthermore, the present legal and judicial climate is so humanistic that court decisions, no matter how strong the evidence and how valid the constitutional position, might very likely go against the creationists. Even in the event of a favorable court decision, the creationists would be bound to lose the case in the biased reporting of the news media. Finally such laws would be very difficult to enforce, even if upheld by the courts. Teachers compelled to teach creationism against their will, and without any adequate knowledge of the creationist arguments and evidence, would probably do more harm than good in the classroom anyway.

So, although the route of persuasion seems slower than that of compulsion, it holds more promise of ultimate success, and our ICR literature has stressed this repeatedly. However, many creationists have felt otherwise and have tried to use the courts or legislatures to get the two-model approach accepted.

This situation has, of course, placed ICR in a difficult position. While not favoring legislative or political action at all, poorly-drawn bills and political defeats would be so harmful that ICR has become inadvertently involved in these activities to try to prevent damaging errors. For example, a model creation *resolution* (not *law*) was prepared by ICR in order to enable school boards to *encourage* (not *compel*) a two-model approach, and ICR scientists and attorneys have been allowed to go on temporary leave in order to serve as expert witnesses or deputized attorneys in connection with creationist litigation. The costs of all such activities have been borne not by ICR but by the organizations requesting them.

The Arkansas creation law resulted, of course, in a lawsuit filed by the ACLU and, finally, a strongly negative decision striking down the bill. Even more hurtfully, it resulted in a news media circus and a great wave of bad publicity for the whole creationist cause. At this writing (September, 1982) a similar ACLU lawsuit against a somewhat similar law in Louisiana is scheduled for trial sometime in mid-1983.

In spite of the bad precedent set in Arkansas, one good result may have been accomplished. The tremendous publicity generated by the Arkansas trial, coming on top of the increasing interest generated by the creationist books, debates, seminars, and other activities, has caused a sharp upturn in public awareness of the issue, and apparently even a significant increase of belief in creationism.

A Gallup poll in 1980 showed that over half the population of the United States believes in a literal, specially-created Adam and Eve as the parents of the whole human race. Then a 1981 Associated Press-NBC News poll found that no less than 76% of the people wanted creation to be taught along with evolution in the public schools, and that another 10% wanted *only* creation to be taught! Then, in August 1982, another Gallup poll found that 44% of the people believe not only in creation, but in *recent* creation, less than 10,000 years ago. Thirty-eight percent *more* believed in God as Creator, though they believed in an old earth and a divinely-guided process of evolution. Only 9% believed in atheistic evolution and 9% were undecided.

These sampling statistics are especially significant in light of the many decades of indoctrination in the schools, textbooks, and news media to the effect that the earth is billions of years old, total evolution is a scientific fact, and creationism, being purely religious, has no place in the schools. No wonder the evolutionists are now beginning to become almost hysterical in their opposition to the creationist movement.

An interesting comment appeared following the Sacramento lawsuit of early 1981, filed by a creationist group (the Creation Science Research Center) against the State on the basis that exclusive evolutionary teaching in the schools was an infringement on their religious rights.

> For his own part, (attorney for the CSRC) Turner says his recent experience in the courtroom has whetted his desire for more. "These scientists get up on the stand, and act as if their very lives were being attacked. They not only close ranks, but they almost deny anybody the right to

know the internal fights that go on within
the evolutionary crowd. They're pompous
and arrogant, just the kind of people that
the First Amendment was written to
protect us against."[1]

From personal experience in many debates,
and with evolutionist questioners (and often
hecklers) in many campus audiences, ICR
scientists can unequivocally echo Mr. Turner's
observations. If evolution is really so scientific
and objectively factual, as evolutionists have
tried to persuade themselves, it is strange that
they quickly become so emotional and angry
when any scientific question is raised about its
validity. These are not appropriate attitudes
and reactions for *scientists!*

The fact is, of course, that evolution is not a
science at all, but a religion, and creationists
are questioning the tenets of a fanatical
religion. No wonder its devotees become
hysterical.

The writer has discussed the essentially
religious character of evolutionism in many
other places,[2] so will not do so here, except to
note that evolutionists themselves are
becoming very sensitive on this issue.

A rallying of the ranks would definitely
be needed if creationists argued that
evolution was a religion. Constitutional

1. Wm. J. Broad, "Creationists Limit Scope of Evolution
    Case," *Science*, Vol. 211, March 20, 1981, p. 1332.
2. For example, *King of Creation* (San Diego: Creation-Life
    Publishers, 1980), pp. 67-108.

scholars do not scoff at the issue, one
expert at Harvard recently saying that it is
"far from a frivolous argument."[1]

The evolutionary religion, of course, is simply humanism (at least as taught in American schools), and even the American Humanist Association acknowledges that humanism is "a non-theistic religion."

> The creationists have portrayed
> Darwinism as a cornerstone of "secular
> humanism," a term they use to describe
> the belief that man, not God, is the source
> of right and wrong. They blame humanist
> teaching for all sorts of modern ills—from
> juvenile delinquency to the high rate of
> abortions—and want to replace it with the
> teaching of Christian morality . . . . As the
> creationists' goals become clear, many
> scientists, realizing that they have been
> secular humanists all along, are beginning
> to marshal their forces . . . . Evolutionists
> are beginning to realize that, for the first
> time in half a century, they may have to
> defend themselves in court.[2]

Some sense of the alarm in the evolutionists' camp is indicated by the following excerpts from a remarkable fund-raising letter recently circulated by the atheistic humanist Isaac Asimov, on behalf of the American Civil Liberties Union and its coming Louisiana creation lawsuit:

---

1. William J. Broad, op cit, p. 1331.
2. Joel Gurin, "The Creationist Revival," *The Sciences*, New York Academy of Science, Vol. 23, April 1981, p. 34.

> We must be prepared for the long and costly battle of challenging every creationist statute in every state in which it is introduced. Unbelievable as it may seem, there are millions of Americans who call themselves "scientific creationists." These religious zealots . . . are marching like an army of the night into our public schools with their Bibles held high . . . . Today, I am writing my personal check for $100 to help the ACLU finance this important case. I urge you too to write your check today. Your help is needed desperately right now.[1]

As the author of nearly 250 books, many of them bestsellers, plus an almost innumerable number of popular magazine articles, Dr. Asimov will probably be able to afford his contribution, but it does sound like his blood pressure may need monitoring.

Another interesting commentary on the state of mind of evolutionists today is provided by Norman Macbeth. After first discussing the article refuting natural selection, written by Professor Ronald Brady (already noted in Chapter 2, p. 43), Macbeth narrates the following:

> A few minutes ago I mentioned Ron Brady's article on natural selection in *Systematic Zoology*. I will not name the

---

1. Isaac Asimov, Fund Appeal Letter for American Civil Liberties Union, March 1982, 4 pp.

man or the college in this case, but it was an Ivy League College and a respectable man . . . . So they told (a student) to go on down to the library . . . and read it right there. He came back in half an hour and said, " . . . the article isn't there, it's been scissored out." Next day the assistant professor went into the office of the head of the department on some other business and on the head's table he saw the missing pages . . . . The head of the department said, "Well, of course I don't believe in censorship in any form, but I just couldn't bear the idea of my students reading that article." End of story.[1]

Macbeth, who is on close speaking terms with many leading evolutionists, especially in the northeastern states, says that they are almost irrationally fearful of the creationists today and are determined to stop them by any means possible. Furthermore:

> . . . they are not revealing all the dirt under the rug in their approach to the public. There is a feeling that they ought to keep back the worst so that their public reputation would not suffer and the Creationists wouldn't get any ammunition.[2]

1. Norman Macbeth, "Darwinsim: A Time for Funerals," *Towards,* Vol. 2, Spring 1982, p. 22.
2. *Ibid.*

THE WAR AGAINST CREATION

Time and space do not allow us to survey the multitude of anti-creationist books and articles being published today. Practically every scientific and educational journal, most of the popular magazines, and most of the nation's newspapers have featured one or more anti-creationist articles in the past few years. Practically all are characterized by gross misunderstanding at best and blatant falsehood at worst, presenting a distorted caricature of creationism and the creationist movement in general and often attacking ICR and its scientific staff in particular. The same false charges (long since answered) are repeated over and over, apparently on the principle that wishful thinking will make them so. A large number of both local and national telecasts, as well as radio broadcasts, have likewise attempted to smear the creationists. Always the theme is that creationists are nothing but a small group of religious fundamentalists attempting to force their unscientific views on other people, while evolutionists are careful, sober-minded scientists.

But this media attack has not worked, and more and more people are becoming creationists. Not only the American Civil Liberties Union, but many other organizations are becoming more and more active in anti-creationist propaganda. The most prestigious of all scientific organizations, the National Academy of Sciences, hosted a meeting in October 1981, with representatives from many

key organizations[1] present, to plan a broad
nationwide anti-creationist strategy for the
years ahead.

The meeting produced many suggestions. A
paper circulated among the participating
societies by A. G. Lasen, Executive Director of
the Assembly of Life Sciences, National
Academy of Sciences, after the meeting,
presented a summary of its recommendations.
Among these were the following:

> . . . a communications network called
> Committees of Correspondence . . . to use
> political action at the local level.[2]

These local committees had already been
initiated by an activist group in the AAAS, and
they have indeed become quite active in
fighting creationism at the local level.
Continuing with other recommendations:

> It was generally agreed that debates
> were to be avoided . . . . However, it was
> recognized that debates might be

---

1. Among organizations included were the American
    Association for Advancement of Science, American
    Humanist Association, National Association of Biology
    Teachers, American Anthropological Association,
    National Science Teachers Association, American
    Museum of Natural History, Smithsonian Institution,
    American Society of Biological Chemists, National
    Cancer Institute, Biological Sciences Curriculum Study,
    American Geological Institute, American Institute of
    Biological Sciences, and others.
2. Alvin G. Lazen, "Summary Report: Meeting on
    Creationism-Evolutionism," *National Academy of
    Sciences*, Washington, D.C., October 19, 1981, p. 2.

> unavoidable, . . . and that it was necessary
> to find means to identify appropriate
> debaters and to better prepare them for
> their task . . . . It was suggested by several
> persons that an "Institute for Evolution
> Research" was needed to counter the San
> Diego-based Institute for Creation
> Research.[1]

There was much more, but a key paper was that presented by veteran creationist-fighter John A. Moore, of the University of California at Riverside. Dr. Moore's long diatribe included at least a dozen major recommendations, many of them involving substantial costs. But first he made a very significant acknowledgment.

> The climate of the times suggests that
> the problem will be with us for a very long
> time . . . .[2]

One thing is sure—the situation will never revert back to the way it was before, as evolutionists might wish. There are now thousands of fully qualified scientists who have become creationists and many more thousands of creationist students in the universities and colleges. Almost half the American population now believes in special, recent creation, notwithstanding decades of evolutionary brainwashing in the schools, and almost 90%

---

1. *Ibid*, p. 3.
2. John A. Moore, "Countering the Creationists," paper presented to the National Academy of Sciences Ad Hoc Committee on Creationism, Washington, D.C., October 19, 1981, p. 1.

want creation to be brought back into the public schools and taught as a viable model of origins. The evolutionists have good cause to be concerned—the creationists are no longer an insignificant fringe minority.

Dr. Moore goes on to recommend that a national coalition of societies and universities be formed to fight creation. This national consortium would, among other things:

(1) Establish a national information-gathering network
(2) Keep the scientific and educational establishments informed
(3) Assemble anti-creationist statements and classroom materials
(4) Maintain a list of key individuals for different assignments
(5) Support travel and other costs of anti-creationist projects
(6) Secure cooperation of liberal church groups, social scientists, and others
(7) Collect funds from foundations and individuals
(8) Conduct anti-creationist short courses for teachers and others
(9) Reform scientific education to pre-empt creationist inroads
(10) Contact lawmakers and other people of key influence
(11) Encourage evolutionists to speak with force and authority
(12) Seek to persuade every American scientist and science teacher to contribute about $10 annually for the work of the consortium

Dr. Moore concluded his paper with the

following cogent claim (unfortunately a true claim):

> If we do not resolve our problems with the creationists, we have only ourselves to blame. Let's remember, the greatest resource of all is available to us—the educational system of the nation.[1]

There have been a number of similar "war councils" and other meetings sponsored by various organizations in the last two years. A group of scientists and others associated with the American Humanist Association have started publishing a regular anti-creationist journal and the National Biology Teachers Association publishes a regular anti-creationist newsletter. At least ten bitter and misleading anti-creationist books have either just been published or are in process of publication as of this writing (September 1982).

This horrendously costly and troublesome anti-creationist vendetta is, of course, all so unnecessary. Since evolution is presumably a fact of science, all the evolutionists need to do is to present one scientific *proof* of evolution. Or even some real unequivocal scientific evidence. *That* would stop the creationist movement cold! Why haven't they thought of that?

---

1.   *Ibid,* p. 6.

Chapter 6

Christian Evolution and Flaming Snowflakes

EVOLUTION AND THE COMPROMISING CHRISTIAN

The modern creationist revival has caused great distress to another important group. Ever since Darwin, a significant number of Christian clergymen and Christian intellectuals have felt it essential to be in line with "modern" thought and have therefore tried to accommodate evolution in their theology. Many of the greatest preachers of the nineteenth century— such as Henry Drummond in England and Henry Ward Beecher in America—soon were leading large numbers of their followers into theistic evolution, and it wasn't long before practically all the seminaries and colleges of the mainline denominations had capitulated to what they considered to be modern science.

The fundamentalist revivals of the early twentieth century did result in the establishment of a number of conservative denominations, as well as non-denominational schools, and these all tried to oppose evolution, along with other modernistic beliefs. However, the Scopes trial of 1925 seemed to discourage a positive creationist stand even among the evangelical and fundamentalist schools and churches. For the next quarter century, most of the Christian leadership in these groups compromised to one degree or another with the evolutionary system. Some frankly accepted theistic evolution, many more went to the day-age theory and probably even more advocated the gap theory, but almost all accepted the evolutionary age-system.

Now, Christians who compromise rarely realize or recognize that it is compromise. With all good intentions they are trying to win people to Christ, and to do this, they feel they must not make the Christian faith seem anti-intellectual or the Bible unscientific. Since, they assume, science had "proved" evolution to be true and—even more clearly—the earth to be very old, it is necessary to use some system of Biblical exegesis to accommodate the vast evolutionary ages of geology and biology in the Genesis record.

These currents of seemingly well-meaning compromise soon became a swelling tide with the rise of Christian intellectualism and the neo-evangelical movement of the forties and fifties. The many evangelical and funda-mentalist colleges established in the early years of the century had by mid-century gone far down the same path of compromise already traveled by the Christian schools of the previous century.

It was in this context that the modern creationist movement began. At first it was not overtly trying to confront the vast secular evolutionary establishment, but primarily seeking merely to call Bible-believing Christians back to faith in true creationism as taught in the Bible. This new generation of creationist scientists sought first of all to show their own Christian brethren that the real facts of science did not require the various exegetical compromises into which they had been led, but rather that true science supported a straight-forward literal exegesis of the Bible. Many Christian responded gladly, and creationism began to thrive again.

But many resisted. Many Christian intellectuals in the universities, as well as many scholarly seminarians, did not wish to face the prospect of alienating the academic world in which they were becoming more and more comfortable, and they bitterly resented the growing Christian awareness that their position was one of compromise. They desired to retain both the position of honor accorded them by their Christian brethren on Sunday and the academic prestige of the community of scholars on Monday, and the creation movement was a growing threat to this tenuous posture.

As a result, the neo-evangelical scientists and theologians have opposed strict creationism almost as vigorously as the evolutionary humanists whose favor they curry. They commonly seek to rationalize this opposition by spiritual arguments, of course. Creationists are guilty of "polarizing" Christians instead of fostering Christian unity, they say. Also such "unscientific" arguments may turn many young people away from Christianity altogether when they find that the scientific community as a whole repudiates these arguments and is irrevocably committed to the "fact" of evolution. It is safer to hold to a *modified* doctrine of "creation," they think, than to risk the loss of all these young people who will be unable to hold to the strict creationism required by a literal reading of Genesis. And, even if Christians still want to believe in a few acts of special creation, they certainly should not question the evolutionary geological-age system as a whole, nor maintain that the Genesis flood was a worldwide cataclysm, as the Bible implies, since this would make them look ridiculous to modern geologists.

CHRISTIANS AND THE UNCOMPROMISING EVOLUTIONIST

But the tragic thing about such compromises is that they never work! One compromise leads to another, and the evoluionary establishment is never satisfied with anything less than total atheism. The road of compromise is always a one-way street which finally ends in a precipice. The doctrine of creation simply cannot be separated from the doctrine of salvation. If creation goes, Christianity goes. As the atheist Bozarth has written:

Christianity has fought, still fights, and

will fight science to the desperate end over evolution, because evolution destroys utterly and finally the very reason Jesus' earthly life was supposedly made necessary. Destroy Adam and Even and the original sin, and in the rubble you will find the sorry remains of the son of God. Take away the meaning of his death. If Jesus was not the redeemer who died for our sins, and this is what evolution means, then Christianity is nothing.[1]

It is significant that atheists seem to understand this fact better than many Christians. Note also this evaluation by one of the world's leading philosophers.

In cultures such as ours, religion is very often an alien form of life to intellectuals. Living, as we do, in a post-Enlightenment era, it is difficult for us to take religion seriously. The very concepts seem fantastic to us . . . . That people in our age can believe that they have had a personal encounter with God, that they could believe that they have experienced conversion through a "mystical experience of God," so that they are born again in the Holy Spirit, is something that attests to human irrationality and a lack of a sense of reality.[2]

---

1. G. Richard Bozarth, "The Meaning of Evolution, *American Atheist,* September 1978, p. 30.
2. Kai Nielsen, "Religiosity and Powerlessness," *The Humanist,* Vol. XXXVII, May-June 1977, p. 46. Nielsen is Professor of Philosophy at the University of Calgary.

Thus, if Christians think that a compromise with evolution will be all that is necessary, and that they can still retain their faith in the deity and redemptive work of Christ, and in "personal Christianity," they are in for an eventual rude awakening. As Bozarth says:

> What all this means is that Christianity cannot lose the Genesis account of creation . . . and get along. The battle must be waged, for Christianity is fighting for its very life . . . . If we work for the American Atheists today, Atheism will be ready to fill the void of Christianity's demise when science and evolution triumph.[1]

Note the undercurrent of intended future forcible suppression of religion in Nielsen's comments. Those who believe in conversion and the new birth through the Holy Spirit are "irrational" and "lack a sense of reality." According to the 20th century's leading evolutionist, Sir Julian Huxley, this is true even of those who believe in God:

> Darwinism removed the whole idea of God as the creator of organisms from the sphere of rational discussion . . . . we can dismiss entirely all idea of a supernatural overriding mind being responsible for the evolutionary process.[2]

---

1. G. R. Bozarth, op cit, p. 30.
2. Julian Huxley, in Issues in Evolution (Ed. by Sol Tax, University of Chicago Press, 1960), p. 45. Huxley was the keynote speaker at the great Darwinian Centennial Convocation at the University of Chicago in 1959.

Thus, the *doctrines* of Christianity are "nothing," if evolution is true and the *experiences* of Christianity are "irrational," if evolution is true. It is with such opinions and philosophies that "Christian evolutionists" feel they must compromise. The esteem in which Bible-believing Christians are held by the anti-religious community is vividly displayed on the cover of a recent issue of the *American Atheist:*

> American religious fundamentalism is not only an increasingly malignant threat to human dignity, intellect, and reason, but also one of the most calculated campaigns of demogoguery, hate, cruelty, greed, ignorance, persecution, intolerance, oppression, injustice, exploitation, and pseudo-Christian barbarity that has existed within the borders of a civilized nation.[1]

And, lest anyone think that such opinions are characteristic only of the rather crude atheism promulgated by Madalyn Murray O'Hair's American Atheist organization, consider that expressed by Dr. Edward Wilson of Harvard University:

> Bitter experience has taught us that fundamentalist religion, which in its aggressive form is one of the unmitigated

---

1. Attributed to Randell Thomas (1980), and quoted in bold letters on the front cover of the August 1982, Vol. 24, issue of *American Atheist,* together with a large swastika and the caption: "Moral Majority: the Way of the Cross."

> evils of the world, cannot be quickly
> replaced by benign skepticism and a
> purely humanistic world view, even among
> educated and well-meaning people . . . .
> Liberal theology can serve as a buffer.[1]

Wilson is one of the world's leading entomologists and is founder and leader of the growing field of so-called "sociobiology." His hatred of fundamentalism may be directed especially at the Southern Baptists, of which he was once a member, and that of the American Atheists directed especially at Jerry Falwell and the Moral Majority, but all such writers think that any Christian who believes the Bible to be God's Word is a fundamentalist and that this is a bad thing to be.

Neo-evangelicals are inclined to be as opposed to Falwell-type fundamentalism as they are to strict creationism, but they should realize that this will not in any wise endear them to the doctrinaire evolutionists. The latter may *use* them, but they won't *accept* them. They should note carefully Wilson's comment that, although full-blown humanism cannot be imposed *quickly* on society, liberal theology can be used as a "buffer"!

---

1. Edward O. Wilson, "The Relation of Science to Theology," *Zygon*, Sept./Dec. 1980. This was in a paper presented at the 1979 Star Island Conference, cosponsored by the American Academy of Arts and Sciences, and the Institute of Religion in an Age of Science.

THE NON-CHRISTIAN NATURE OF EVOLUTION IN ACTION

"But," many Christians say, "Can't a Christian also be an evolutionist? Why can't we regard evolution as God's method of creation? Why do you want us to choose between evolution and creation. Why can't we believe in both?"

Well, as a matter of fact, one *can* be a Christian evolutionist. One *can* be a Christian liar or a Christian thief or a Christian gossip, too. Christians can, unfortunately, be many things they ought *not* to be! This author himself was a Christian evolutionist until some time after getting out of college, but that didn't make it right. The Bible clearly teaches special creation—not evolution— and so did the Lord Jesus Christ Himself. There can be, and are, some Christian evolutionists, but there is no such thing as "Christian evolution." A Christian is one who accepts Jesus Christ as Savior and Lord, submitting to His authority and believing His Word. Since Christ taught the literal historicity of the Genesis record (e.g., Matthew 19:3-6; Luke 17:26-29), so also the *consistent* Christian must do.

"Christian evolution," by its very nature, is thus inconsistent in concept, a contradiction in terms, like "Christian atheism" or "flaming snowflakes" or "kindly cruelty." Christian evolutionists may, indeed, be born again, saved by grace, redeemed by the saving work of Christ on the cross, but they will also one day have to explain to the Lord just why they chose to believe the false science of evolution instead of the plain statements of God's Word.

The philosophy of evolution is a substitute for God, not an attribute of God. It does not help people come to God, but rather leads them away from Him. The testimony of Edward Wilson, the world's leading sociobiologist, is illuminating in this regard:

> As were many persons in Alabama, I was a born-again Christian. When I was fifteen, I entered the Southern Baptist Church with great fervor and interest in the fundamentalist religion; I left at seventeen when I got to the University of Alabama and heard about evolutionary theory.[1]

This testimony could be duplicated times without number. Charles Darwin himself started out as an orthodox creationist Christian, but was eventually led by his own evolutionary philosophy into agnosticism and atheism. Dr. Huston Smith, a leading philosopher and also Professor of Religion at Syracuse University, comments on this phenomenon as follows:

> One reason education undoes belief is its teaching of evolution; Darwin's own drift from orthodoxy to agnosticism was symptomatic. Martin Lings is probably right in saying that "more cases of loss of religious faith are to be traced to the theory of evolution . . . than to anything

---

1. E. O. Wilson, "Toward a Humanistic Biology," *The Humanist*, September/October 1982, p. 40.

else." (*Studies in Comparative Religion,*
Winter 1970).[1]

Nor does it prevent such loss of faith by
trying to tell young people that the Bible
accommodates evolution, for anyone who can
read the Bible will quickly discover this is not
the case. Ten times in the very first chapter of
Genesis we are told that God created each kind
of plant and animal to reproduce only after its
own kind. At the end of that first chapter, we
are told that God *stopped* creating and making
things, so that processes going on now are *not*
creative processes at all. In modern
terminology, processes going on now cannot
evolve anything new. There may be much
horizontal variation, within the kinds, but never
any *vertical* variation, from a lower kind to a
higher kind.

A great many additional reasons[2] can be
cited to show that a harmonization of evolution
with the Bible is quite impossible. Perhaps the
most succinct and powerful of all is the
statement of God in the Ten Commandments:
"Remember the sabbath day to keep it holy. Six
days shalt thou labour, and do all thy work: But
the seventh day is the sabbath of the Lord thy
God: in it thou shalt not do any work, . . . : For

---

1. Huston Smith, "Evolution and Evolutionism" *Christian
    Century,* July 7-14, 1982, p. 755. This journal is the
    semi-official spokesman for religious liberalism.
2. See, for example, *Scientific Creationism* (Gen. Edn., Ed.
    by Henry M. Morris, Creation-Life Publishers, San Diego,
    1974), pp. 214-247.

in six days the Lord made heaven and earth, the sea, and all that in them is, and rested the seventh day: wherefore the Lord blessed the sabbath day and hallowed it" (Exodus 20:8-11).

The Ten Commandments, of course, mean little to an atheist or to anyone else who does not believe in the God of the Bible. For the Christian, however, one would suppose that these commandments—written with the very finger of God (Exodus 31:18)—would be taken seriously, and literally, if *any* part of the Bible would be so taken. And this Commandment says as clearly as could possibly be expressed in human language, that God's work-week was precisely the same as man's work-week. In fact, the one was given as a specific pattern and justification for the other. If God's "days" were not the same as man's "days," then we have been deliberately deceived by God! And that right in the middle of the most sacred and inviolate of all His divine revelations, the only one which was not only divinely *inspired,* but also divinely *inscribed!*

How, then can a conscientious Christian possibly submit to evolution in the face of such a clear declaration by his Maker! Furthermore, the progressive creation idea, with its day/age theory—as well as the gap theory, with the evolutionary ages pigeonholed in an imaginary gap before Genesis 1:2—are both likewise precluded by this simple and comprehensive statement of the Creator's.

Thoroughgoing evolutionists—the ones who are the leaders of evolutionary thought and best understand the implications of evolution— certainly are not impressed by this compromising posture of the theistic evolutionist. Jacques Monod, for example, Nobel Prize-winning microbiologist and certainly one of Europe's top evolutionists, said shortly before his death:

> If we believe in a Creator . . . it is basically for moral reasons, in order to see a goal for our own lives. And why would God have to have chosen this extremely complex and difficult mechanism. When I would say by definition He was at liberty to choose other mechanisms, why would He have to start with simple molecules? Why not create man right away, as of course classical religions believed?[1]

This is the same question creationist Christians have been presenting to theistic evolutionists for years, only to be impatiently turned away by irrelevant rejoinders. Maybe they will listen to it when it is posed by an atheistic biologist such as Monod.

But Monod also echoes the creationists with another point, even more of an indictment against so-called Christian evolutionists:

> [Natural] selection is the blindest, and most cruel way of evolving new species, and more and more complex and refined organisms . . . . The struggle for life and elimination of the weakest is a horrible

---

1. Jacques Monod, "The Secret of Life." transcript of television interview with Laurie John, on the Australian Broadcasting Co., June 10, 1976. Reprinted in *Ex Nihilo* (Journal of Creation Science Association of Australia).

process, against which our whole modern
ethics revolts. An ideal society is a non-
selective society, one where the weak is
protected; which is exactly the reverse of
the so-called natural law. I am surprised
that a Christian would defend the idea
that this is the process which God more or
less set up in order to have evolution.[1]

When an atheistic, socialistic humanist like
Monod expresses surprise at such Christian
compromise, it is time to ask our Christian
evolutionist intellectuals to take a hard and
critical look at their own motives and the
direction they are trying to take us. Monod is at
least honest in his evaluation of evolution and
its true and ultimate implications. In a very
influential book, he said:

Where then shall we find the source of
truth and the moral inspiration for a really
scientific socialist humanism, if not in the
sources of science itself, in the ethic upon
which knowledge is founded, and which by
free choice makes knowledge the supreme
value—the measure and warrant for all
other values?
It prescribes institutions dedicated to
the defense, the extension, the enrichment
of the transcendent kingdom of ideas, of
knowledge, and of creation—a kingdom
which is within man, where progressively
freed both from material constraints and
from the deceitful servitudes of animism,
he could at last live authentically,

1.   *Ibid.*

protected by institutions which, seeing in him the subject of the kingdom and at the same time its creator, could be designed to serve him in his unique and precious essence.[1]

This eloquent eulogy exalts *knowledge,* or *science* as the measure of all value in the coming socialistic and humanistic world kingdom, with man as its creator. But what about the promised kingdom of God? Monod answers:

The ancient covenant is in pieces; man knows at last that he is alone in the universe's unfeeling immensity, out of which he emerged only by chance. His destiny is nowhere spelled out, nor is his duty. The kingdom above or the darkness below: it is for him to choose.[2]

Although it was certainly not the intended meaning of this final sentence of his widely influential book, Jacques Monod has actually issued an appropriate challenge to the Christian evolutionist, strikingly echoing Elijah's ancient challenge to the prophets of Baal: "How long halt ye between two opinions? If the Lord be God, follow Him; but if Baal, then follow him" (I Kings 18:21).

An evaluation of theistic evolution by the great British evolutionary biologist, Peter Medawar, also a Nobel prize winner, is similar

---

1. Jacques Monod, *Chance and Necessity* (New York: Alfred A. Knopf, 1971), p. 180.
2. *Ibid.*

to that of Monod. He has little sympathy for
anyone who professes to see evidence of divine
guidance of the evolutionary process:

> Unfortunately, the testimony of Design
> is only for those who, secure in their
> beliefs already, are in no need of
> confirmation. This is just as well, for there
> is no theological comfort in the ampliation
> of DNA, and it is no use looking to
> evolution: the balance sheet of evolution
> has so closely written a debit column of all
> the blood and pain that goes into the
> natural process that not even the
> smoothest accountancy can make the
> transaction seem morally solvent
> according to any standards of morals that
> human beings are accustomed to.[1]

It is, indeed, a cause for perplexed wonder
that so many evangelical Christians, who
profess to believe the Bible and in a God of
wisdom, power, and love, could also believe in
evolution, which is so completely inconsistent
with the very concept of a wise, powerful, and
loving God. Evolution is the most foolish,
wasteful, and cruel process that could ever be
invented by which to produce man. Surely an
omniscient God could *devise* a better way, an
omnipotent God is *capable* of a better way, and
a *loving* God would certainly *prefer* a better
way, than evolution. If one wants to believe in

---

1. P. B. Medawar and J. S. Medawar, *The Life Science:
    Current Ideas of Biology* (New York: Harper and Row,
    1977), p. 169.

evolution for some reason, that is his affair, but he should not blame God for it. The God described in the Bible would never be guilty of such a thing!

Not only is the very concept of evolution inimical to true Christianity, but so are the effects. Evolutionists often express resentment at the creationist charge that evolutionary philosophy has been made the basis of so many evil social systems—such as Communism, Fascism, racism, imperialism, social Darwinism, ethical relativism, behaviorism, anarchism, etc., etc. While acknowledging that the founders and promoters of these systems have, indeed, used evolution as the intellectual rationale for their systems, modern evolutionists insist that evolutionary theory has been misunderstood and misapplied in such cases. In like manner, they say, Christian doctrine has been misapplied in many wicked actions perpetrated in the name of Christianity—the Inquisition, the religious wars, the witch hunts, etc. Evolution should not be blamed for Nazism any more than Christianity should be blamed for the Crusades, they say.

However, the fact is that evolutionary theory has really *not* been misapplied in such systems. It does, by its very nature, eliminate the need for God and any absolute standard of righteousness and moral judgment. Only that which survives can evolve, and so survival necessarily is the chief good. In the ultimate sense, might *does* make right, and the weak and unfruitful *should* be eliminated.

The great Will Durant, one of the world's

leading humanists, certainly neither Christian nor creationist, has called attention to this fundamentally anti-moral nature of evolutionism:

> By offering evolution in place of God as a cause of history, Darwin removed the theological basis of the moral code of Christendom. And the moral code that has no fear of God is very shaky. That's the condition we are in . . . . I don't think man is capable yet of managing social order and individual decency without fear of some supernatural being overlooking him and able to punish him.[1]

Even though Durant himself felt no personal need of God (or so he would have us think), he believed that a stable social and moral order required belief in God. This conclusion of course implies that widespread belief in evolution would lead to an unstable—even chaotic—state of society and morality. He therefore felt we are in great need of a revival of genuine faith in God.

> I should say we are now in the last stage of the pagan period and that consequently I would expect there would be a resurrection of religious belief and an aid to the moral life that is an aid to civiliza-tion . . . . Chaos is the mother of dictator-ship, and then the sequence goes all over again. Dictatorship is the mother of order,

---

1. Will Durant, "We are in the Last Stage of a Pagan Peroid," *Chicago Tribune Syndicate,* April 1980.

order is the mother of liberty, liberty is the mother of chaos, chaos is the mother of dictatorship, and then the cycle renews.[1]

DID GOD REALLY MEAN WHAT HE SAID?

This discussion could be extended at great length, but what has already been written should more than suffice to demonstrate the complete incompatibility of evolution with theism in general and Christianity in particular.

However, much of the opposition in the Christian world to modern creationism is ostensibly not against creation *per se* but against *recent* creation. Many evangelical scholars in recent years have published books and articles in which they claim to believe in creation, while at the same time they are very anxious to defend the geological-age system and the great antiquity of the earth. The anti-creationist hysteria of the evolutionists is thus now being echoed even by the progressive creationists.

For that matter, much of the anti-creationist literature from the evolutionary establishment also centers on these two points. Evolutionists are well aware that their *specific* evidences for evolution are very weak, even though they maintain that the *general* evidence favors evolution. They find it much easier to attack the concepts of a young earth and flood geology than to present concrete scientific

1. *Ibid.*

evidence for evolution.

Creationists have pointed out repeatedly that scientific creationism can be taught quite independently of Biblical creationism, and that this is the only form of creationism requested for public institutions. Furthermore, geological catastrophism can be taught independently of Biblical catastrophism, and the *scientific* evidence for recent creation can be taught with no reference whatever to the *Biblical* evidence for a recent creation.

In fact, all three of these issues—creationism versus evolutionism, catastrophism versus uniformitarianism, and young earth versus old earth—are distinct subjects, and it is not at all necessary to mix them together in a classroom or textbook discussion. They do, of course, relate closely to each other and affect each other, but they are nevertheless fundamentally distinct from each other.

For example, it would be perfectly possible for someone to believe in a real period of special creation of all things, say, ten billion years ago, with all processes having operated uniformly since that time. Conversely, one could believe in the recent formation of all things by naturalistic, but cataclysmic, evolutionary processes. An advocate of the first could be called an old-earth uniformitarian creationist, an advocate of the latter a young-earth catastrophic evolutionist. Although neither position is widely held, of course, there is nothing philosophically impossible about either of them, or, for that matter, about any one of several other plausible combinations of the six variables. It is obvious that the reason creationism is usually associated with catastrophism and a young earth is because all three are taught in the Bible, and evolutionists desperately want to tag creationism with the onus of religion. The progressive creationists do have a point in this connection. But even if the Bible is not interpreted in terms of a recent creation and worldwide deluge—as they prefer —the basic issue of creation versus evolution still has to be addressed. Creationists maintain that the real facts of science (e.g., the laws of thermodynamics, the complexity of living organisms, the pervasive fossil gaps) still require special creation even if the earth is old and there never was a flood. And *this* is what needs to be included in tax-supported education.

As far as catastrophism and the young earth are concerned, these are separate issues *scientifically,* and each should be discussed on its own merits. There are many good scientific evidences for both a young earth and a worldwide flood, and there is no good reason why these should not be included in public education, but they still do not need to be tied to the basic creation/evolution issue, as far as public schools are concerned. In this very book, for example, various facets of the modern evolutionist turmoil are discussed, but the current quite intense geological discussions about catastrophism versus uniformitarianism are *not* included, since these only relate peripherally to our basic subject.

On the other hand, we do believe that scientific creationism and Biblical creationism are quite compatible and that the latter *does* relate to Biblical catastrophism and Biblical chronology, since all are components in the great plan of God. We can understand the insistence of the evolutionist that scientific creationism be tied to the flood geology and young earth implications of the Bible. This is the obvious strategy for him to use, since he has no real scientific evidence for evolution. With what he feels is a stronger case for the geological ages and an old earth, on the other hand, he can divert attention from the basic issue by fighting these Biblical teachings and then alleging these to be integral components of scientific creationism. This strategy often seems to work well in debates and popular polemics.

Both Christian evolutionists and progressive creationists, therefore, accept the full geological-age system and join with the secular evolutionists in opposing Biblical creation. They apparently hope thereby to avoid the academic ostracism which strict creationism usually incurs.

Those of us who are committed to recent creation and flood geology, of course, would also much prefer to be honored by our academic colleagues instead of vilified. We get no pleasure out of being slandered and ridiculed. We did earn the same kinds of terminal degrees, have held the same kinds of scientific positions, and have published the same kinds of scientific books and articles, as our evolutionist colleagues. On the surface it would seem to be much easier, much more remunerative, and much more secure merely to try to conform to the evolutionary humanism that dominates the intellectual world.

This is the way it evidently seems to those of our Christian brethren who opt for the old-earth world view of the evolutionist. As a matter of fact, however, the committed evolutionist will not be satisfied with *this* compromise, nor with any other that stops short of total capitulation to atheism, as we have already pointed out.

Furthermore the cost of such compromise is too great. Temporal gain is no compensation for eternal loss. God's Word *does* teach recent creation and a global cataclysm, and the geological ages (whether or not certain acts of creation are allowed in them here and there) do teach an age-long spectacle of waste and suffering and death, thus libelling God's character. This fact has already been established earlier in this chapter and is elaborated at great length in many other creationist books. Similarly there is an abundance of geological and ethnological evidence for a recent worldwide hydraulic cataclysm (and, therefore, for flood geology), as well as other physical evidences for a recent creation. That is not the subject of this book, however, and the interested reader is directed to many of the ICR books and monographs[1] for these evidences.

The recent Gallup poll on this subject, mentioned in the previous chapter, has been

---

1. A complete descriptive book list is available on request to the Institute for Creation Research, P.O. Box 2666, El Cajon, California 92021.

most illuminating and encouraging in this connection. The following report summarizes the salient findings of this significant poll.

> The American public is almost entirely divided between those who believe that God created man at one time in the last 10,000 years and those who believe in evolution or an evolutionary process involving God.
>
> Of the participants in the poll, 44 percent, nearly a quarter of whom were college graduates, said they accepted the statement that "God created man pretty much in his present form at one time within the last 10,000 years."
>
> Nine percent agreed with the statement: "Man has developed over millions of years from less-advanced forms of life. God had no part in this process."
>
> Thirty-eight percent said they agreed with the suggestion that "man has developed over millions of years from less-advanced forms of life, but God guided this process, including man's creation."
>
> Nine percent of those interviewed simply said they did not know.[1]

These results were quite unexpected and produced predictable expressions of shock from the array of evolutionary scientists and liberal theologians who were asked to comment. The surprising thing, of course, was

---

1.   Gallup Poll, "44% Believe God Created Mankind 10,000 Years Ago," *San Diego Union,* August 30, 1982. From *New York Times Service.*

that almost half the population believes not only in creation, but in *recent* creation! And yet evolutionists—including our "Christian evolutionists"—have been ridiculing this notion for years, saying that nobody except a lunatic fringe of Biblical fundamentalists could believe such a thing. Note also that 25% of these believers (about the same number as in the general population) were college graduates.

Now where can we suppose all these millions of people in our population ever could have learned this idea? They certainly did not get it from the schools in which they studied (especially all those college graduates!), nor did they get it from the news media which bombard them daily with evolutionary propaganda of various degrees of subtlety.

There seem to be only two basic sources from which they could ever have learned this strange doctrine of a recent creation. The literature of modern scientific creationism has apparently been convincing to many who have read it, in spite of the bombastic rebuttals to this literature by evolutionists and in spite of the textbooks they had studied in their schools. Most believers in recent creation, however, no doubt learned this truth simply by reading and believing the Bible. The various distortions of Genesis designed to accommodate the evolutionary geological ages have apparently been recognized as such and rejected by most readers of the Bible, neo-evangelical intellectuals to the contrary notwithstanding.

Unfortunately, as of this writing (September 1982), probably the majority of evangelical leaders are still fearful of really believing the

Bible on this foundational issue, and thus are
still rationalizing various ways of getting
around God's Word when it affirms that He
made all things in six literal days (Exodus
20:8-11) and that the antediluvian world was
destroyed in a global flood (II Peter 3:3-5). On
the other hand, multitudes of rank-and-file
Christians, including great numbers of young
college graduates (along with a significant and
growing number of evangelical scientists and
theologians) have already returned to their
fathers' faith in the literal historicity and
perspicuity of the Biblical record of creation
and earth history.

Chapter 7

The Way of Peace

PEACE OR TURMOIL

To the Christian creationist, the turmoil within the evolutionary and humanistic establishments today can hardly fail to call to mind all the Biblical passages which predict such confusion for all who refuse God's peace. The following passages are typical:

> For thus saith the high and lofty One that inhabiteth eternity, whose name is Holy; I dwell in the high and holy place, with him also that is of a contrite and humble spirit, to revive the spirit of the humble, and to revive the heart of the contrite ones . . . . I create the fruit of the lips; Peace, peace to him that is far off and to him that is near, saith the Lord; and I will heal him. But the wicked are like the troubled sea, when it cannot rest, whose waters cast up mire and dirt. There is no peace, saith my God, to the wicked (Isaiah 57:15, 19-21).
>
> And the way of peace have they not known: There is no fear of God before their eyes (Romans 3:17-18).

Lest anyone misunderstand, the term "wicked" as used in Scripture, although it may refer in various contexts to any number of different specific "sins," is primarily defined as in the last verse cited above. That is, the wicked are those who have no fear of God, and who, therefore, can never know the way of peace. In the Old Testament, the same Hebrew word for "wicked" is translated "ungodly" (e.g., Psalm 1:1). Psalm 9:17 says that "the wicked shall be turned into hell, all the nations that forget God," thus defining *wickedness* also as "forgetting God." Satan himself, the arch rebel against the word of God and the authority of God, is called "that Wicked One" (e.g., Matthew 13:18-19; 37-39).

The clear teaching of Scripture is that those who "forget God," those who have "no fear of God," those who deny the word of God, are "wicked" and shall therefore have no peace. That such characterizations apply to the leaders of evolutionary thought (and, by and large to their followers, as well) is surely evident from the documentation provided in the previous chapters of this book. Although this philosophy of ungodliness may appear outwardly stable and powerful, it inevitably must lead to turmoil and eventual disintegration. As the psalmist David, inspired by the Holy Spirit said, long ago:

> I have seen the wicked in great power, and spreading himself like a green bay tree. Yet he passed away, and, lo, he was not: yea, I sought him, but he could not be found. Mark the perfect man, and behold

the upright: for the end of that man is peace (Psalm 37:35-37).

THE TROUBLED SOUL OF CHARLES DARWIN

The life and death of Charles Darwin himself is a case in point. Although he has received the adulation of the world for over a century, and was buried with high honors in Westminster Abbey, he never knew peace himself after once placing his faith in evolution. As his latest biographer, Irving Stone, has noted:

> Darwin returned to England at 27 in a robust state of mind and body. It was not until a year later, when he began to write in his evolutionary notebooks, that he first felt and commented on his illnesses, forcing himself into a lifetime of severe, repugnant, and sometimes ludicrous disability.[1]

Darwin complained all of his life of his constant illnesses, and entire books have been written just about this aspect of his life. Yet in the most exhaustive and modern volume on this subject, a book entitled *To Be An Invalid*, the author, Dr. Ralph Cox, concludes that there was nothing organically or physically wrong with him at all. His granddaughter, Nora Barlow,

---

1. Irving Stone, "The Death of Darwin," Chapter 22, in *Darwin Up to Date* (Ed. by Jeremy Cherfas, London: *New Scientist* Guide, IPC Magazines, Ltd., 1982), p. 69. Stone is author of the best selling new biography of Darwin entitled: *The Origin: A Biographical Novel of Charles Darwin*.

who edited his autobiography, said he was a hypochondriac.

The biographer Irving Stone, who is an ardent evolutionary humanist and profound admirer of Charles Darwin, attributed all his troubles to the intense conflicts generated by his evolutionary theory, blaming opposition of the Christians and creationists of the day. Stone does acknowledge, however, that Darwin hated to "think about the demon of evolution he had released upon an unwilling and unprepared world"[1]

Whatever the cause, Charles Darwin was a vigorous, healthy, almost happy-go-lucky young man before he was converted to evolution, but a man of sickly body and troubled mind all his life thereafter. Stone also is anxious to repudiate the widely circulated story of Darwin's repentance and conversion during his final days:

> Upon word of his death, his detractors circulated a rumor that he had repented on his deathbed, and asked God's forgiveness for his blasphemies. There was not an iota of truth to the charge, yet it still surfaces today, presented as fact by those who would like to believe it.[2]

---

1. *Ibid.*
2. *Op cit*, pp. 69-70. Although there is certainly no firm evidence of Darwin's reputed conversion, there are some possible intimations. See Malcolm Bowden, *The Rise of the Evolution Fraud* (San Diego: Creation-Life Publishers, 1982), pp. 188-193.

It is not surprising, of course, that belief in evolution leads eventually to inward conviction of guilt, and outward conflict and turmoil. If God does, indeed, exist, and we are, indeed, His creatures (and this is surely the teaching of the Bible and all *true* science), then our very minds and hearts are bound to be programmed God-ward. Rebellion against God—whether in terms of philosophical denial, active disobedience or careless neglect—is bound, therefore, ultimately to deprive mind and heart and body of the spiritual sustenance they require from their offended Creator.

THE CREATOR OF PEACE

But if the broad road of evolutionism leads to turmoil, how does one who has been following that road ever find "the way of peace?"

The answer, of course, is through the Creator, who is none other than the Lord Jesus Christ. He is "the Way, the Truth, the Life" (John 14:6). The mission of His forerunner, John the Baptist, was said specifically to be "to guide our feet into the way of peace" (Luke 1:79). He is the "Prince of Peace" (Isaiah 9:6), and His gospel is the "gospel of peace" (Ephesians 6:15). Those who have come to him in repentance and faith (this writer included) know the experience of "joy and peace in believing" (Romans 15:13), and "being justified by faith, we have peace with God through our Lord Jesus Christ" (Romans 5:1).

That Christ is personal Creator, as well as personal Savior, is the clear teaching of the New Testament (John 1:1-3, 10; Ephesians 3:9; Hebrews 1:2; etc.). The great passage in

Colossians 1:16-20 presents the Lord Jesus in His great three-fold work as Creator, Sustainer, and Redeemer of all things. Note the grand scope of His work, from eternity to eternity—- past, present, and future:

(1) *Past Work.* "For by Him were all things created, . . . : all things were created by Him, and for Him;; (Colossians 1:16).

(2) *Present Work.* "And He is before all things, and by Him all things consist" (Colossians 1:17).

(3) *Future Work.* "And, having made peace through the blood of His cross, by Him to reconcile all things unto Himself" (Colossians 1:20).

Note, especially from verses 16 and 20, in their whole contexts, that these mighty works encompass everything in heaven and earth. Note also that His work of creation was completed in the past and is not continuing in the present, as theistic evolutionists would have it. Note further that His present work is that of conservation (Greek *sunistano,* "sustain," "consist"), reflected scientifically in the universal law of conservation of energy and matter. Note finally that His future work, that of reconciling all things back to Himself, is assured on the basis of the redemption price of His blood shed on the cross, and that the end of it all is *peace!*

"Peace" is a wonderful word and peace has been an age-long dream of men and nations. There have been peace conferences, peace treaties, peace parties, peace marches, peace symbols, peace prizes—yet real peace seems always elusive and ephemeral. It is one thing to talk about peace, but how do we *make* peace?

The answer is here in Colossians 1:20. Jesus "made peace with the blood of His cross." The reason why there is no peace between man and man is because there is no peace between man and God. Conflicts of nation against nation, class against class, man against man, husband against wife—conflicts of all kinds at all levels—are merely symptomatic of a deadly cancer in the human bloodstream. "From whence come wars and fightings among you? Come they not hence, even of your lusts that war in your members?" (James 4:1).

The *sin* innate in human nature is not an evolutionary product of millions of years of animal ancestry, as humanists believe, but is simply rebellion against God and His Word, inherited through the rebellion of our first parents. God created man "in his own image" (Genesis 1:27) and placed him, as His steward, in dominion "over every living thing that moveth upon the earth" (Genesis 1:28), but Adam and Eve sinned, rejecting God's Word and believing the great Adversary. This wicked choice severed them from the divine fellowship and resulted in God's curse of decay and death on man and all his dominion (Genesis 3:17-19).

Instead of completely repudiating His rebel creatures, however, as strict justice would warrant, our loving and merciful Creator began a new work on their behalf. "Upholding all things by the word of His power" (Hebrews 1:3), the Lord Jesus Christ initiated the great principle of conservation, or salvation, holding it all together until His full plan of redemption could be completed.

It is well to be reminded that the two greatest laws of science—the universal principles of conservation and decay, the two laws of thermodynamics—are merely the scientific formulations of, first, God's completed and conserved work of creation, and second, His curse on the creation because of sin. These laws were formulated by modern scientists only in the nineteenth century but were, in effect, recorded in Scripture from the very beginning.

The alienation of man from his Creator has continued on through all these long ages since the fall of man. But the price of redemption, as well as assurance of ultimate reconciliation, was finally provided when Christ died on the cross and rose from the dead. The Word became flesh (John 1:14); God became man, and the Creator became the Savior. And there He made peace through the blood of His cross, to reconcile all things unto Himself.

Here is the way of peace! Not conformity, not compromise, not surrender, but sacrificial love in the context of absolute truth and justice. The price of peace is the great love of God, in Jesus Christ. Not only has He created us, with a wonderful creation to enjoy, but He also has died to redeem and reconcile us to Himself, suffering the awful judgment of separation from the Father, in substitution for all the guilty sinners of all the world through all the ages. Then, He arose from the grave, eternally triumphant over sin and death, able therefore "to save them to the uttermost that come unto God by Him, seeing He ever liveth to make intercession for them" (Hebrews 7:25).

Only through our great Creator and Savior,

therefore, lies the way of peace. The infinite love of God in Christ is the standard against which the continuing rebellion of men and women today must be measured. Forgiveness of sin and reconciliation to God the Creator are offered freely to all who will believe and receive Him. "You that were sometime alienated and enemies in your mind by wicked works, yet now hath He reconciled in the body of His flesh through death, to present you holy and unblameable and unreprovable in His sight" (Colossians 1:21-22).

It is *this* measure against which the arrogant unbelief of modern evolutionists must be judged. "Of how much sorer punishment, suppose ye, shall he be thought worthy, who hath trodden under foot the Son of God, and hath counted the blood of the covenant, wherewith he was sanctified, an unholy thing, and hath done despite unto the Spirit of grace? . . . . It is a fearful thing to fall into the hands of the living God" (Hebrews 10:29,31). "How shall we escape, if we neglect so great salvation?" (Hebrews 2:3).

No doubt some who read these words—such as the evolutionist spokesmen who have taken it upon themselves to monitor the writings of the creationists—will react either with anger or sarcasm, considering all of this merely as further excuse for excluding scientific creationism from public institutions (notwithstanding the fact that, as they well know, creationists themselves do not want these doctrines of Christianity even mentioned in public schools).

On the other hand, it is the hope and prayer

of the writer and his associates that many who read the book will be those students and others who are really searching for truth, and for peace of soul and confidence of mind. They may have been (as almost everyone today has been) influenced and confused by the pervasive teachings of evolutionary humanism in our schools and news media and are now looking for a better way.

We trust such readers will see that evolutionism is both false and harmful, and then will enter, by faith, the way of peace, trusting the Lord Jesus Christ as Creator, Savior, and Lord, for now and for ever.

Indexes

Index of Names

A

Ager, Derek V. 66
Asimov, Isaac 35, 132, 133
Ayala, Francisco J. 48, 88

B

Badham, Harry 19
Barlow, Nora 169
Beecher, Henry Ward 141
Bethell, Tom 112
Bird, Wendell R. 125
Bismarck, Otto 92
Blyth, Edward 38
Boudinot, Elias 124
Bowden, Malcolm 170
Bozarth, G. Richard 144,
  145, 146
Britten, Roy J. 59
Broad, William J. 131, 132
Busse, P. H. 80

C

Carnegie, Andrew 92
Cherfas, Jeremy 169

Clemmey, Harry 19
Conner, Cliff 99
Cox, Ralph 169
Cracraft, Joel 118, 119,
  120, 121, 123
Crick, Frances 32

D

Darwin, Charles 16, 38, 39,
  43, 45, 99, 150, 158, 169,
  170
de Santillana, Giorgio 77
Dobzhansky, Theodosius
  44, 91
Down, John L. H. 94, 95
Drummond, Henry 141
Dubay, Dennis 118
Durant, Will 157, 158

E

Eigen, Manfred 23, 24, 25,
  26
Eiseley, Loren 38
Eldredge, Niles 63, 83, 84,
  91, 100, 101, 102
Engels, Frederick 109, 110

F

Falwell, Jerry   148
Feduccia, Alan   65
Fleischer, Gerald   69
Fox, Sydney   34
Franz, Peter   39

G

Gardiner, William   23
Gish, Duane   111
Goldschmidt, Richard   85,
   86
Goodman, Morris   59
Gould, Stephen J.   30, 31,
   38, 39, 40, 42, 63, 65, 73,
   83, 84, 85, 86, 91, 94, 95,
   96, 100, 101, 102, 103,
   104
Grassé, Pierre P.   49, 50, 55
Gribbin, John   17, 18, 20
Gurin, Joel   123

H

Haeckel, Ernst   92, 96
Haldane, J. B. S.   44
Halstead, Beverly   109, 110,
   111
Heikes, K. E.   80
Herbert, W.   80
Hitler, Adolph   92, 98
Hoffman, P.   98
Hofstadter, Douglas R.   21
Hopson, James P.   78
Hoyle, Fred   26, 27, 29, 32,
   33, 34
Huxley, Julian   44, 91, 146

J

Jefferson, Thomas   125
Jepsen, Glen L.   44
Johanson, Carl Donald   74,
   76, 104
John, Laurie   153

K

Kammerer, Paul   100
Kemp, Tom   68
Kimura, Motoo   48
Koestler, Arthur   41, 42, 43
Kojata, Gina Bari   55
Korey, Kenneth A.   56, 57,
   58

L

Lasen, Alvin G.   136
Laszlo, Ervin   108
Leakey, Louis   81
Leakey, Mary   79
Leakey, Richard   74, 76, 80,
   81
Leek, Brian   39
Lepkowski, Wil   108
Lewin, Roger   51, 54, 68
Lings, Martin   150
Lombard, R. Eric   69
Lysenko, Trofim   100

M

Macbeth, Norman   43, 133,
   134
Mann, Alan   77

Marx, Jean L. 65
Marx, Karl 98, 99, 112
Mayr, Ernst 44, 91
Medawar, Jean B. 156
Medawar, Peter B. 155
Mendel, Gregor 50
Miller, Stanley 17, 30
Monod, Jacques 153, 154, 155
Moore, John A. 137, 138
Morris, Henry M. 11, 94, 106, 116, 123, 131, 151
Morris, John D. 82
Mussolini, Benito 98

N

Nelson, Gareth 82
Nielsen, Kai 145, 146
Nietzsche, Friedrich 92

O

O'Hair, Madalyn Murray 147
Olson, Storrs L. 64
Oparin, Alexander 16
Orgel, Leslie 22, 32
O'Rourke, J. E. 102
Osborn, Henry Fairfield 96, 97
Oxnard, Charles 78, 79

P

Paley, William 38
Parker, Gary E. 106
Pasteur, Louis 16

Patterson, Colin 39, 42, 55, 56, 119
Pavlov, Ivan 100
Perlas, Nicky 45, 46
Pesely, Gregory Alan 41
Pilbeam, David 73
Platnick, Norman 82
Popper, Karl 110
Prigogine, Ilya 107, 108, 109

R

Radinsky, Leonard B. 78
Raup, David M. 42, 75
Rhodes, Cecil 92
Ridley, Mark 61, 62
Rockefeller, John D. 92

S

Sagan, Carl 104
Sarich, Vincent 58
Schindewolf, O. H. 85, 86
Schopf, L. J. M. 42
Schuster, Peter 23
Schwabenthan, Sabine 96
Sepkoski, J. 42
Simberloff, D. S. 42
Simpson, George Gaylord 40, 44, 91
Smith, Huston 150, 151
Spencer, Herbert 92
Stanley, Steven M. 57, 70, 84, 85, 86 91, 100
Stebbins, G. Ledyard 44, 88, 91
Stone, Irving 169, 170

T

Tashian, J. H.   59
Tashian, R. E.   59
Tax, Sol   146
Thomas, Randell   147
Turner, Richard K.   130, 131
Tuttle, Russell   80

V

von Dechend, Hertha   77

W

Waddington, C. H.   40
Watson, Lyall   73

Whitcomb, John C.   116
Wickramasinghe, Chandra 32, 33, 45
Williams, Emmett L.   106
Wilson, A. C.   55, 58
Wilson, Edward O.   147, 148, 150
Winkler-Oswatitsch, Ruthild 23
Wright, Sewall   44, 87, 88

Y

Young, Robert M.   103

Z

Zuckerman, Solly   78

Index of Subjects

A

Adaptation 37-38
  See also: Natural selection
American Academy of Arts
  and Sciences 148
American Association for
  Advancement of Science
  45, 118, 136
American Civil Liberties
  Union 118, 129, 132,
  133, 135, 136
American Humanist
  Association 132, 136,
  139
American Museum of
  Natural History 119, 136
Anti-creationism
  Atheistic opposition 144-
  146
  Evolutionary establish-
  ment 135-139
  Neo-evangelicals 143-
  144, 148, 159, 162
  Opposition to Biblical
  Christianity 147-148,
  159
  See also: Evolutionism

Archaeopteryx 63-64, 66
Atheism
  Battle against creationism
  144-146
  Darwinism 16, 145
  Humanism 132
  Impossible to prove
  34-35
Atmosphere, Primeval
  16-20
Australopithecus 75, 78-81

B

Baupläne 62, 63, 64, 85
Bible, Teachings of
  Christ 171-172
  Creationism 122, 124,
  127, 151-152, 163
  Flood geology 166
  Folly of atheism 35
  Genesis kinds 63, 151
  No peace for ungodly
  167-168, 171, 175
  Recent creation 151-152,
  165-166
Big Bang theory 106

Birds, Evolution of 63-66
Blood proteins   54-55

C

Catastrophism
    Cosmic   105-106
    Dissipative structures
        107-109
    Flood geology   160, 163
    Naturalistic geology   105
    Punctuated equilibrium
        105-106
    Revolutionism   109
Christ
    Creator   89, 171, 172
    Creationism, Teacher of
        149
    Savior   145, 149, 171,
        172
    Sustainer   172
Christian evolution
    Contradiction in terms
        149
    Harmful to Christian faith
        150-151
    Inconsistent with God's
        character 152-159
    Repudiated by Scripture
        151-152, 172
    See also: Theistic
        evolution
Church and State, Separa-
    tion of 122, 125-126
Circular reasoning
    Natural selection   37,
        40-43
    Paleontology   102

Cladistics   42, 81-82
Communism
    See: Marxism
Compromise with evolution
    by Christians
    Christian evolution   149-
        152
    Day-age theory   93, 115,
        142, 152
    Gap theory   93, 115, 142,
        152
    Immorality of evolutionary
        concept   152-159
    Liberal theology   114, 148
    Progressive creationism
        159, 161, 162
    Theistic evolution   141-
        143
    Uncompromising atheistic
        evolutionism   144-148,
        155
Conservation principle
    First law of thermody-
        namics   174
    Natural selection   38-39
Constitution, United States
    125-126, 127
Creation Research Society
    116, 127
Creation Science Research
    Center   130
Creationism
    Americanism, Foundation
        of   124-126
    Biblical   122-123
    Evidence from science
        33-34, 89, 120-121
    Evolutionist concern
        45-46, 133-134

Modern Revival   113-115,
   143
Opposition   122-123,
   135-139
Young-earth question
   130, 159-166
Cro-Magnon Man   78
Cytochrome c   54, 55

D

Darwinian Centennial   45,
   92, 146
Darwinism
   See: Mutations; Natural
   Selection; Neo-Darwinism
Day-Age theory   93, 115,
   142, 152
Debates, Creation/Evolu-
   tion   11, 111, 116-121,
   131
Directed panspermia   32
Dissipative structures   107-
   109
DNA   21-22, 51, 53-54
Drosophila   49

E

Earth, Age of
   See: Young-earth
   creationism
Environmental Mutagenic
   Society   46-47
Escherichia coli   49
Evolutionism
   Fossil evidence   61, 64,
   67-71

Gradualism   63, 65, 83,
   91, 111
Internal conflicts   12-13,
   94, 100, 109-113
Man   72-83
Mechanism unknown   52,
   94, 113
Opposition to Christian
   theism   114, 141-143,
   149-159
Pervasiveness in all fields
   114-115
Punctuated equilibrium
   63, 83-85
Religious nature   131-132
Revolutionary   83, 100-
   104, 109
Stasis   84
Thermodynamic barrier
   106-109
Extraterrestrial life   32-33

F

Fetoscopy   95-96
Flood geology   160, 161,
   163
Footprints, Fossil hominid
   79, 80, 81
Fossils
   Evolutionary evidence   61
   Hominoid   72-83
   Missing transitions   62-
   72, 76, 83-84, 121
   Mosaic organisms   64, 66,
   68
   Recapitulation theory   95,
   96

G

Gap theory   93, 115, 142, 152
Genetics
  Coding for proteins   54
  Diversity of gene types   51
  DNA similarities   53-54
  Genetic code origin   21-22
  Mendelism   50
Geology
  Catastrophism   105-106
  Circular reasoning   102
  Flood geology   160, 163
  Gradualism   63, 65, 83, 91, 111

H

Heidelberg Man   77
Hemoglobin   52, 54-55
Hominid fossils
  Australopithecus   75, 78-81
  Confusion among anthro-
    pologists 76-77
  Fragmentary nature   72-73
  Homo erectus   80-81
  Repudiated   77-78
  Homo erectus   80, 81
  Homo habilis   78
Hopeful monsters
  Concept originated and
    rejected   85
  Concept recently revived
    86
Impossibility of concept   87-88
Horse evolution   69-71
Human origins
  See: Hominid fossils; Man
Humanism   132
  See also: Atheism;
    Evolutionism
Hyracotherium   70

I

Insects   48
Institute for Creation
  Research
  Debates   116-120
  Graduate school   117
  Non-political   127-129
  Origin and organization
    116-117
  Publications   11, 117, 163
  Research   117, 137

J

Java Man   77

K

Kinds of organisms
  Bauplane   62, 63, 64, 85
  Genesis   63, 64, 65, 85
  Stability   83-85

L

Laetoli footprints   79, 80
Lamarckianism   16, 99-100
Legal problems in creation
    movement   127-133
Liberal theology   114, 148
Life, Origin of
    Creation   33-34
    Early in earth history
        30-32
    Laboratory experiments
        17, 34
    Naturalistic explanation
        essential for evolution
        15
    Outer space   32-33
    Oxidizing atmosphere
        18-20
    Primeval soup   16-18
    Probability barrier   28-30,
        33
    Reducing atmosphere
        needed   16-19
    Thermodynamic barrier
        22-25, 34, 106-109
Lucy   78, 80, 104

M

Mammal-like reptiles   67-69
Man, Evolution of
    Australopithecines   75,
        78-81
    Chimpanzee similarities
        53-54, 58
    Cranial size variations   75
    Disagreement among
        anthropologists   76-77
    Fragmentary nature of
        evidence   72-73
    Primates, No fossil links to
        72
Marxism
    Atheistic character   99
    Evolutionary basis   99,
        103
    Inadvertent alignment
        with creationism   112
    Infiltration in education
        92-93
    Neo-Darwinian reaction
        109-111
    Prestige schools   103
    Punctuationism   102-104
    Rapid evolution required
        100-101
Molecular clock
    Evolutionary sequences
        53
    Inconsistencies   56-58
Mosaic organisms   64, 66,
    68
Moth, Peppered   48
Mutations
    Beneficial, rare or non-
        existent   47-48, 50
    Eliminated by natural
        selection   39
    Error process   47
    Harmful   44, 46, 85, 107
    Miracles required   50
    Monsters   85

N

National Academy of
    Sciences   135, 136, 137

National Biology Teachers
    Association   136, 139
Natural selection
    Conservation principle
    38-39
    Inability to create new
    kinds   38-40
    Just-so stories   42, 70
    Tautologous   37, 40-43
Nazism   92, 98, 159
Neanderthal Man   78
Nebraska Man   77
Neo-Darwinism
    Evolutionary synthesis   44
    Laissez-faire capitalism
    92, 98
    Modern questioning   45
    Mutations   47-50
    Natural selection 37-43
    Older generation of
    evolutionists   91
    Racism   94-98
    Resistance to punctua-
    tionism   87-88, 109-111
    Social Darwinism   92, 98
    Sociobiology   111
Neo-evangelicalism   114,
    142, 143

O

Origin of life
    See: Life, Origin of

P

Peace
    Darwin's troubled soul

    169-170
    Promised to believers
    167-169, 171, 172-173,
    174-176
    Provided by Christ   171,
    173, 176
    Rebellion and sin prevent
    167-168, 171, 173, 175
Peking Man   77
Piltdown Man   77
Polls, Creation-Evolution
    Associated Press   12, 129
    Gallup   12, 129, 130,
    163-165
Population genetics   56-57
Probability   28-30, 33
Progressive creationism
    93, 115, 142, 152, 159,
    161, 162
Proteins
    Divergence rates   56-57
    Origin   25
    Supposed evolutionary
    sequences   53
Punctuated equilibrium
    Catastrophism   105-106
    Fossil gaps   63
    Neo-Darwinian opposition
    66
    Origin of life   31-32
    Russian background   83,
    100-102
    Student movements of
    sixties   91-94

Q

Quantum speciation   85, 87

R

Racism
  Darwinism basis 94-98
  Modern evolutionary
    racists 96-97
  Not derived from Scripture
    97
Ramapithecus 78
Recapitulation theory
  94-96
Recent creation
  See: Young-earth
    creationism
Revolutionary evolutionism
  83, 100-104, 109
Rhodesia Man 77

S

Schools, Teaching in
  Creationist background
    124-125
  Legal battles 127-129,
    130, 132-133
  Two-model approach
    123-124, 127, 128
Scopes trial 93, 113, 142
Second law of thermody-
  namics
  Barrier to evolution
    22-25, 34, 106-107
  Biblical curse 174
  Dissipative structures
    107-109
  Open system argument
    27-29
Similarities
  Common designer 53

Supposed evidence of
  evolution 53-54
Skull 1470 78
Social Darwinism 92, 98
Socialism
  Humanism 154
  Marxism 101, 109
  Nazism 98, 157
Sociobiology 111, 148, 150
Spontaneous generation 16
Stasis 49, 84
Stereo-chemistry 25
Systematics 56

T

Theistic evolution 114,
  141-143, 172
  See also: Christian
    evolution
Transitional forms in
  evolution, Supposed
  Archaeopteryx 63-64, 66
  Hominid forms 72-83
  Horse series 74, 76
  Mammal-like reptiles
    67-69
  Mosaic forms 64, 66, 68
  No real transitional forms
    62-72, 76, 83-34, 121
Two-model approach in
  education 123-124, 127-
  128

Y

Young-earth creationism
  Biblical teaching 165-166

Gallup poll support   12,
    130, 164-165
Neo-evangelical opposition
    159, 162
Physical evidence   163
Scientific question distinct

from creation-evolution
issue   160-162

Z

Zinjanthropus   78

The Genesis Record *Henry M. Morris, Ph.D.*
Verse by verse scientific and devotional commentary on the book
of beginnings. **No. 070, Cloth**

Dinosaurs: Those Terrible Lizards *Duane T. Gish, Ph.D.*
Did dinosaurs live at the same time that humans did? Are
dragons just imaginary creatures? In this beautifully illustrated
book for children, Dr. Gish explains what dinosaurs were and why
they no longer exist. The issue of creation vs. evolution is
presented on a level that children can easily understand.
No. 046, Cloth

Dry Bones . . . and Other Fossils *Gary E. Parker, M.S., Ed.D.*
What are fossils? How are they formed? What can we learn from
them? These and many other questions are answered in conversa-
tional dialogue in this creatively illustrated book for children. Its
educational value is enhanced by including references to the fall
of Adam, the Flood, and the promise of a new earth. An explanation of fossils and their significance presented in a manner that
children will understand and enjoy. **No. 047**

The Natural Sciences Know Nothing of Evolution
A. E. Wilder-Smith, Ph.D.
Examines the evidence and presents the conclusions in a com-
prehensive analysis of evolution from the viewpoint of the Natural
Sciences. **No. 110**

Occult Shock and Psychic Forces
Clifford Wilson, Ph.D., and John Weldon, M.A.
The relatively obvious occult involvements have expanded to include more subtle forms of deception—parapsychology, hyp-
nosis, est, acupuncture, and holistic medicine. This activity is
rampant—you need to be aware of it. You need to read this timely
book. **No. 113**

Has Anybody *Really* Seen Noah's Ark? *Violet M. Cummings*
A documented report of the search for Noah's Ark that reads like
a detective story. Facts that have come to light in the last few
years, as well as a recap of the history of the search. A fascinating
section on the Michigan Mounds. 75 photos, 416 pages. **No. 072**